Vascularization in Tissue Engineering

Edited by

Xiaoxiao Cai

West China School of Stomatology
Sichuan University
China

Vascularization in Tissue Engineering

Editor: Xiaoxiao Cai

ISBN (Online): 978-981-14-7584-9

ISBN (Print): 978-981-14-7582-5

ISBN (Paperback): 978-981-14-7583-2

© 2020, Bentham Books imprint.

Published by Bentham Science Publishers Pte. Ltd. Singapore. All Rights Reserved.

need for a court order if at any point you breach any terms of this License Agreement. In no event will any delay or failure by Bentham Science Publishers in enforcing your compliance with this License Agreement constitute a waiver of any of its rights.

3. You acknowledge that you have read this License Agreement, and agree to be bound by its terms and conditions. To the extent that any other terms and conditions presented on any website of Bentham Science Publishers conflict with, or are inconsistent with, the terms and conditions set out in this License Agreement, you acknowledge that the terms and conditions set out in this License Agreement shall prevail.

Bentham Science Publishers Pte. Ltd.
80 Robinson Road #02-00
Singapore 068898
Singapore
Email: subscriptions@benthamscience.net

CONTENTS

PREFACE

Angiogenesis refers to the process of forming a new vascular system with an existing vascular network through the proliferation and migration of vascular endothelial cells on the basis of existing capillaries and/or venules. Angiogenesis is of great significance in tissue engineering, wound healing and regeneration repair. The core of tissue engineering is to establish a three-dimensional space complex of cells and biological materials, which is used to reconstruct the morphology, structure and function of the damaged tissues and achieve permanent replacement, to achieve the purpose of repairing wounds and rebuilding functions. However, there are few tissue engineering products that can be applied to clinics at present. One of the main reasons is the early vascularization of tissue engineering products. Evidence shows that when the cell mass is greater than 3 mm^3, the diffusion of interstitial fluid cannot be used to maintain the cell's survival, and the supply of oxygen and nutrients must be achieved through the regeneration of blood vessels. Because tissue engineering scaffolds do not have a vascular network, vascular regeneration has become an important factor limiting the ability of tissue engineering constructs to form tissues after implantation. Therefore, when using tissue engineering technology to construct large and complex artificial tissues or organs, how to avoid ischemic necrosis or poor healing of the central part of the defect has become the primary task of tissue engineering vascularization. At present, by constructing scaffold material-seed cell-growth factor complex, based on the research foundation of angiogenesis mechanism, the formation of neovascularization *in vitro* three-dimensional microenvironment or *in vivo* transplantation culture is the most common method for tissue engineering vascularization. Therefore, it is very important to understand the occurrence, development, physiological and pathological processes of angiogenesis for tissue engineering vascularization. In this book, the authors focus on the biological and pathological conditions of vascularization, microenvironment factors on angiogenesis, co-culture systems and scaffold materials used for angiogenesis, including: (1) biological basis of vascularization; (2) effects of microenvironment factors on angiogenesis; (3) microenvironment of pathological vascularization; (4) vascularization in co-culture systems; (5) vascularization and scaffold material.

Though this book, readers will have a better understanding of the occurrence, development, physiological and pathological processes of angiogenesis, and know more about ways and means of angiogenesis. The authors sincerely hope that this book will add further insight into basic and applied researchers as well as clinicians involved in tissue engineering vascularization, thus contributing to further advances in regenerative medicine.

Xiaoxiao Cai
Professor of West China School of Stomatology
Sichuan University
Vice-director of Dental Implant Center
West China Hospital of Stomatology
No. 14, 3rd Sec
Ren Min Nan Road
Chengdu 610041
P.R. China

FOREWORD

When Dr. Cai asked me to write this foreword, I was greatly honored to have the opportunity to introduce this eBook.

As one of the research hotspots of life science in the 21ˢᵗ century, tissue engineering is a new subject combining cell biology and material science to construct tissue or organs *in vitro* or *in vivo*. The method is to plant seed cells on scaffold materials to form an active complex, which finally establishes tissues and organs with normal physiological functions and is applied to the treatment and repair of human diseases. At present, the construction of skin, bone and cartilage in vitro has made great progress, and has been widely used in clinical. However, how to build a blood supply system that can transport nutrients, oxygen and remove metabolize waste for the body, to ensure the survival of implanted artificial tissue in the body and play its normal physiological function, which is a major difficulty of tissue engineering, as well as a key issue of regenerative medicine.

Vascularization, existing in both physiological and pathological procedures, is a considerable complicated process which is regulated and controlled by a variety of biological factors via diverse pathways and mechanism. It's a complex and multi-step physiological process, which must be carried out under a strict micro-environment. At present, in addition to conventional treatment methods, plenty of vascular diseases such as arterial ischemic disease, peripheral arterial disease, pulmonary hypertension, limb ischemia, myocardial infarction and cerebral infarction were treated with endothelial progenitor cells possessing the characteristics of stem cells in clinical work, which have a very broad prospect. Therefore, the exploration of the vascularization of tissue engineering will greatly promote the research results of regenerative medicine to clinical application.

It is generally known that seeding cells, growth factors, scaffold materials and microenvironment are the four basic elements of tissue engineering. This eBook is a comprehensive and systematic introduction to the vascularization of tissue engineering from these four aspects. Firstly, it details the whole process of physical vascularization (vascularization and angiogenesis) and the role of various growth factors. Then, the effect of the physiological microenvironment and pathological microenvironment on the vascularization in tissue engineering is elaborated, such as related mechanism, pathophysiological features and so on. Besides that, various common seeding cells, typical co-culture system model and multifarious scaffold materials of vascularization in tissue engineering are also introduced in detail.

In conclusion, I am excited about this eBook. Because it not only systematically and comprehensively introduces the mature theoretical knowledge, methods and technology of vascularization in tissue engineering, but also shows the latest progress achievements and future development in the world. I believe it will be beneficial to all those who have an interest in vascularization in tissue engineering and will lay a crucial and solid theoretical foundation for making future progress of tissue engineering vascularization.

Yunfeng Lin
Professor of Department of Oral and Maxillofacial Surgery
West China College of Stomatology
Sichuan University
Vice-director of State Key Laboratory of Oral Diseases
Director of Laboratory for Nucleic Acid Nanomaterials, SKLOD
Executive Editor-in-Chief of Bone Research
Deputy Editor-in-Chief of Cell Proliferation
China

List of Contributors

Changyue Xue	State Key Laboratory of Oral Diseases, Chengdu, China
Mei Zhang	State Key Laboratory of Oral Diseases, Chengdu, China
Tianyi Zhang	State Key Laboratory of Oral Diseases, Chengdu, China
Xiaoxiao Cai	State Key Laboratory of Oral Diseases, Chengdu, China
Xin Qin	State Key Laboratory of Oral Diseases, Chengdu, China
Yichen Ge	State Key Laboratory of Oral Diseases, Chengdu, China
Yuting Wen	State Key Laboratory of Oral Diseases, Chengdu, China

CHAPTER 1

Biological Basis of Vascularization

Yichen Ge, Yuting Wen and **Xiaoxiao Cai**[*]

State Key Laboratory of Oral Diseases, West China Hospital of Stomatology, Sichuan University, Chengdu 610041, China

Abstract: Vascularization, existing in both physiological and pathological procedures, is a considerable complicated process which is regulated and controlled by a variety of biological factors *via* diverse pathways and mechanism. It is crucial to understand these processes and factors well if we intend to unveil the mystery of vascularization. More importantly, understanding basic processes offer tools and thinking directions with which researchers are able to design various methods to achieve vascularization, which is called vascular tissue engineering. In this part, major procedures of physical vascularization (vasculogenesis and angiogenesis) and growth factors are introduced, as well as their roles and new research outcomes in vascularization. These factors include vascular endothelial growth factor(VEGF), basic fibroblast growth factor(bFGF), platelet-derived growth factor(PDGF), angiopoietin1, angiopoietin2 and others(junctional molecules, integrals *etc.*). Lastly, some frequently used markers and testing methods about vascularization research are briefly introduced.

Keywords: Angiopoietin, Angiogenesis, FGF, PDGF, Vascularization, VEGF.

1. INTRODUCTION OF BIOLOGICAL VASCULARIZATION

It is generally acknowledged that the mechanism of neovascularization comprises two aspects---vasculogenesis and angiogenesis [1, 2]. Supposed as the main mechanism of the construction of vascular networks in the embryonic stage, vasculogenesis occurs in the bone marrow as endothelial progenitor cells (EPCs) gradually migrating, differentiating and finally reforming new vessels, besides it also exists in adult tissue especially in the ischemic area [3, 4]. In the embryo, the first vascular network is built when somites beginning to form by the process of vasculogenesis. Locating between two germ layers, the first blood islands which form the inner layer of the yolk sac occur by *in situ* differentiation from the extra-embryonic mesoderm. In the embryonic stage, vascular network remodeling is characterized as the change of number and/or location of vascular segments to

[*] **Corresponding author Xiaoxiao Cai:** Sichuan University, West China School of Stomatology, China; E-mail: xcai@scu.edu.cn

improve functional adaptability without evident network expansion. Vascular fusion can reduce the number of vascular segments and therefore give rise to larger vessels. Larger vessels in other regions are remodeled into a network of smaller vessels which subsequently increase the number of vessels segments. These biological processes result in a big transformation of primary plexus to enter into a more complexly structured secondary stage. Further expansion of primary and secondary vascular plexus during postnatal life occurs with the process of angiogenesis. Angiogenesis generally refers to the process of new blood vessels formation based on pre-existing ones (Shown as Fig. **1a-c**). The process of angiogenesis comprises two distinct mechanisms, sprouting of endothelial cells (ECs) and splitting of vessel lumens by intussusceptive microvascular growth (IMG).

1.1. Vasculogenesis

1.1.1. Embryonic Vasculogenesis

Vasculogenesis is a process in which blood vessels differentiate from *in situ* into ECs. Initially, the term was used for embryonic development and the presence of angioblasts, which were entirely associated with prenatal growth. "Vasculogenesis" was first brought out by Werner Risau [5] to term the development of mesodermal vascular plexus by differentiation of vascular fibroblasts. Angioblasts are thought to accumulate in the angiogenesis area. Vasculogenesis that occuring embryogenesis and extraembryonic are often accompanied by hematopoiesis, thus the term "angioblasts" has also been proposed. Elementary observations in Sabin [6] strongly suggest that both hematopoictic cells and ECs origins from the angioblasts. The existence of hemangioblasts *in vivo* has developed. At present, instead of defined as an actual ordinary cell precursor, it's rather more like a competitive cell that can generate hematopoietic cells and ECs to local environmental signals [7].

In early mouse embryos, angioblasts originated from decentralized progenitors in the lateral plate mesoderm and expressed Flk-1 (VEGFR-2) and Brachyury (Bry)genes in turn [8]. Bry gene was inhibited and Flk-1 was activated as development progressed. The dynamic changes of these progenitor cell differentiation markers were as follows: Bry +/Flk-1 -, Bry-/ Flk-1 -, Bry +/Flk-1-, Bry +/Flk-1 -, and Bry +/Flk-1 -. These groups represent different phases of differentiation of vascular mother cells. In addition to Flk-1 and Bry genes, they also display diverse sequences of other genic groups [9]. The existence of vascular mother cells has been verified *in vitro* though collecting endothelial and hematopoietic descendants of colony-forming cells (CFCs) derived from embryoid [10, 11]. Embryoid bodies obtained from suspension cultured embryonic stem cells established a model that simulates many cell differentiation in multi-aspect during early somatic embryogenesis [12]. *In vitro*, fast breeding CFCs that produce hematopoietic colonies functioned equally to vascular mother cells. The vascular mother cells have multi-directional differentiation potentials to form ECs, smooth muscle cells and hematopoietic cells under specific conditions [8, 9, 11]. The commitment of angiogenic cells is controlled by transcription factor of etv2/er71gene, the upstream of core genes in the development of EC lineage [13], and activation of endothelial cell line specific genes (endothelin, endothelin and VE cadherin) and hematopoietic and/or hematopoietic (hematopoietic, hematopoietic and SCL) lineages [14]. ETv2 guide differentiation of endothelial and hematopoietic lineage *via* regulating ETS associate genes necessary for downstream stimulation of hematopoietic and endothelial differentiation [14].

Members of the TGF-beta superfamily participated in mesoembryonic expression of Bry, for instance BMP (bone morphogenetic protein) and lymph node and activin signals, as a guide between self-renewal of pluripotent stem cells [15, 16]. Fibroblast Growth Factor-2 (FGF-2) and BMP4 are key signaling ingredients that stimulate the development of embryonic mesoderm, thus accelerate ECs and blood cells production [17, 18]. TFIIS, subtype of TCEA3 which express in mesoderm was proved to drive the production mesoderm EC. As shRNA transfection reduced the expression of TCEA3 in mouse embryonic stem cells, the expression of Bry marker in mesoderm increased, the expression of multipotent genes in mesoderm decreased, the differentiation of EC boosted and the production of vascular endothelial growth factor (VEGF) A increased [19]. Therefore, ECs can bypass a hemangioblast intermediate directly from mesodermal angioblasts. EC differentiation in the period of embryonic process has been demonstrated to be produced by vascular mother cells directly from the mesoderm [20, 21]. These ECs can further form tubules during mesoderm culture *in vitro*. Although current evidence supports the existence of angioblasts, it has been a challenge to isolate these cells and determine their exact location in developing embryos.

Embryonic EC are considered be descendants of angioblasts [22]. Angioblasts were found to transform phenotype in mice: initially expressed tal-1/flk-1, then CD31 was obtained, while the expression of tal-1 was reduced [21]. This phenotypic change of angioblasts was observed during the formation of cardiac ducts, dorsal aortas, interlaminar vessels and main veins in different embryos. During these events, cells processes migration, isolation and assemble into tube/vessel as they differentiate into mature ECs. Instead of classical growth factors such as platelet-derived growth factor (PDGF), VEGF and FGF, notch and ephb2/ephb4 signaling pathways was considered the key components to regulate dorsal aortic angiogenesis [23, 24]. It has been proved that basement-membrane

(BM) and extracellular matrix (ECM) components for instance collagen XVIII, laminin, including beads and sulfate side chains, can dynamically regulate the development of brand blood vessels. The above ECM components are important accumulator for growth factors, which can be fluid into surrounding tissues by selective stimulation thus regulate activity of cell migration, assembly and angiogenesis [25, 26].

1.1.2. Postpartum Vasculogenesis

The discovery of circulating endothelial progenitor cells (CEPC) are a real breakthrough in the field of adult vasculogenesis [27]. The majority of CEPs were bone marrow (BM) derived bone marrow cells or circulating cells of lymphoid lineage, stimulating angiogenesis, and expressing EC markers while stimulating vascular growth factor *in vitro*, but not differentiating into ECs to develop blood vessels [28]. In fact, the term of EPCs has been debated and a new method for selecting these cells has been proposed [29].

Due to the common markers between endothelial progenitor cells (EPC) and hematopoietic cells, there is no suitable method to identify EPC *in vivo* environment, for instance bone marrow (BM), peripheral blood (PB), umbilical cord blood (CB) and solid tissue. The absence of a novel marker, or a collection of markers, and the high uncertainty for isolating and culturing EPCs make it obscure to define these cells. Tissue sources can alter the characteristics of EPCs; when separated from *in vivo* biological cultures, these cells can be influenced by artificial irritation *in vitro*. One of the current methods for identifying EPCs depends on the ability of adhesion and colony *in vitro* with auxiliary usage of flow cytometry techniques to select cells according to their surface phenotypes. Generally speaking, CEPCs screened from mononuclear cells (MNC) origin from peripheral blood are defined as "early" and "late" *in vitro* growth cells. In addition to VEGFR2 (flk-1), the "early" growth cells obtained *via* short-period culture were defined as hematopoietic adhered cells expressing cell surface marker CD14, CD45 and CD11b.They can support angiogenesis and angiogenesis, but cannot spontaneously form tubules *in vitro* [30 - 33]. Early endothelial progenitor cells promote angiogenesis by secreting angiogenesis-promoting molecules in a paracrine manner and can merge into capillaries at perivascular locations [31]. Since proliferation can also be used as a standard for defining these cells, a few groups have constructed a colony formation analysis. Following the initial culture of non-adherent cells for a short-term, EC conditioned colonies were detected; CFCs were also verified to be a origin of medullary lymphoid cells, which stimulated angiogenesis, but could not form tubules [34]. Another type termed "late" endothelial CFCs (ECFCs) are screened following long-term culture *in vitro* up to 2-4 weeks, which do not express any myeloid/lymphoid surface markers such as CD14, CD45 [35, 36]. These cells can directly integrate into the permanent vascular system, and form spontaneous lumen tubules in matrix gel when transplanted internal, and ultimately conduct blood vessels. ECFC is further characterized by colony formation analysis, which means that a different population can be selected, showing high proliferation potential and high telomerase activity [36 - 38].

After pre-screening of CEPCs with magnetic beads enriched CD34/CD133 double positive cells, two different culture groups could be separated. One is termed "primitive EPCs", which forms micro colonies with more proliferation dynamic and stimulated vasculogenesis in postpartum. Preclinical study of human vascular diseases with usage of animal models revealed these primitive EPCs are potential tools for vascular regeneration.

Besides CEPC, there are also a number of resident EPCs. Some permanent EPCs are described as discrete lesions in the endothelium of large vessels, mainly the aorta [39]. Although the conversion rate of EC is relatively low or off the record, there remain some parts in the lining of ECs expressing more proliferation dynamic and high expression of telomerase. Cells situated in these areas are prone to proliferate due to tissue damage. Permanent endothelial progenitor cells were also found in the wall of the artery at the medial edge of the adventitia [40]. These CD34-expressing cells can develop into mature ECs, *via* vascular tubules in transplantable tumors *in vivo*, and promote the formation of tubules in the culture of artery explants. The medial edge of adventitia is the same spot where Sca-1 positive precursor cells was detected. It maintains self-renewal characteristics and can differentiate into other cell lines under selective stimulation, for instance mesenchymal stem/stromal cells (MSCs), osteoblasts and ECs *etc.*

1.1.3. Vasculogenesis in Reparative Process

Current studies have shown that vasculogenesis also happens in adult tissues mostly during wound healing events. The ischemia model revealed that the main source of CEPCs in granulation tissue vasculogenesis is bone marrow(BM). BM-derived progenitor cells can be regulated by SDF-1 (interstitial derived factor), VEGF, erythropoietin, G-CSF (granulocyte colony stimulating factor), statins, bFGF, PLGF (placental growth factor), estrogen, insulin, angiopoietin-1, CXCR4 antagonists and other mediators, such as IL-6 or IL-10 [32, 41, 42]. For example, IL-10 stimulates CEPC present from BM towards wound healing site after myocardial infarction, which seems to be modulated *via* SDF-1/CXCR4 and STAT-3/VEGF signaling pathways.

It has been widely proved that vascular system is distorted and disordered in inflammatory reactions and permanent wound healing events, such as diabetes, obesity, atherosclerosis, hypercholesterolemia, and dyslipidemia (lipid metabolic changes). This is associated with a lower activity and dysfunction of CEPCs, as shown by a lower response to growth factor/chemokine [43].

1.2. Angiogenesis

1.2.1. Introduction of Angiogenesis

Angiogenesis is defined as the development of blood vessels from existing vessels which is important for organ regeneration. A false in this process leads to numerous disorders such as ischemic, infectious, inflammatory and malignant. Blood vessels are evolved to permit hematopoietic cells to examine the body for immune supervision, provide nutrition, and dispose circulating waste. Vessels also provide organ-related indicators in a perfusion-independent manner. Although this process is conducive to tissue differentiation and repair, it can stimulate malignant diseases and inflammation to occur, and is taken advantage by cancer metastasize to kill patients. Blood vessels circulate through every organ, thus abnormal vascular growth can cause many diseases. For example, insufficient growth or maintenance of blood vessels can generate myocardial infarction, apoplectic coma, ulcerative diseases and neural degeneration. Deviating from normal growth of blood vessels can also generate pulmonary hypertension and blinding eye diseases. Several angiogenesis patterns have been verified. In growing mammalian embryos, ECs derived from angioblasts assemble into vascular labyrinths. Different signals participated in vessel differentiation [44]. The following germination generates the production of the vascular network, defined as angiogenesis. Then arteriogenesis starts, and ECs arranged channels are covered by pericytes materials including vascular smooth muscle cells (VSMCs), which provides stability and support perfusion. Blood vessels can also form through other mechanisms, the processes require further studies. For instance, predecessor of blood vessels can be referred to as the process of intussusception splitting, resulting in the production of sub-vessels. In other cases, blood vessels co-selectively occur, in which tumor cells hijack existing blood vessels, or tumor cells can arrange blood vessels development, a phenomenon defined as vascular mimicry. It is hypothesis that stem cells derived from cancer can even generate tumor ECs [45]. Despite the controversy, the assemble of bone marrow-derived cells (BMDC) and endothelial progenitor cells into the vascular border contributes to vascular repair or pathological vasodilation in healthy adults. Progenitor cells then integrate into ECs during postnatal angiogenesis. Collateral vessels bring a large amount of blood flow to ischemic tissues during vascular reconstruction, which is generated by various mechanisms, including attraction and stimulation of bone marrow cells [46].

1.2.2. Branching, Maturation and Resting of Blood Vessels in Angiogenesis

In healthy adults, resting ECs processes a long half-life and are guarded by autocrine controlling signals such as NOTCH, fibroblast growth factor (FGFs), angiopoietin-1 (Ang1) and VEGF. Due to vascular oxygen support, ECs are assembled with oxygen sensors and hypoxia sensible factors, including hypoxia inducible factor 2α (HIF-2α) and prolylhydroxylase domain 2 (PHD2), allowing blood vessels to readjust their morphology to optimize blood circulation. Static ECs form a single layer of phalangeal osteocytes with flow-lined surfaces, which are connected by connecting molecules like claudins and VE-cadherin. ECs are enveloped by pericytes, which release survival signals such as VEGF and Ang1 and inhibit the proliferation of ECs.

When resting vessels sense angiogenesis signals released by hypoxia, inflammation or tumor cells, like Ang2, VEGF-C, FGFs, VEGF and chemokines, pericytes separate first from the vascular wall (the process is in response to Ang2) and isolated away from basement membrane *via* degrading proteolysis regulated by matrix metalloproteinases. VEGF up-regulated the permselectivity of the endothelial cell layer, which led to exosmosis of plasma protein, thus creating an extracellular matrix (ECM) scaffold. ECs managed to transform to the surface of the ECM, *via* integrin signaling. Angiogenic molecules deposited in extracellular matrix was exposed by protease, for instance fibroblast growth factor (FGF) and VEGF, and remodeling the extracellular matrix into a vasoactive environment. In order to prevent ECs from collectively migrating to angiogenesis signals, an endothelial cell, called a type cell, exists such as VEGF receptors, gap ligands DLL4. The adjacent organs of apical cells take stalk cells as auxiliary sites. Regulated by NOTCH, their division lengthens the stalk and stimulated by NOTCH ankyrin repeat protein (NRRAP), WNTs, placental growth factor (PLGF) and FGFs. VE-cadherin, CD34, VEGF and hedgehog mediated the development of lumen [2]. Tip cells are composed of filamentous cells which can sense environmental signals, for instance adrenaline and signal quantities. Stalk cells transmit spatial information about the location of their extracellular environments *via* release of EGFL7 and other elements into the extracellular matrix thus elongate the stalk. During hypoxia process, stimulated by HIF-1A, ECs respond to extracellular angiogenesis signals. Bone marrow scaffold cells assist vessels to fuse to start blood flow. Blood vessels are only functional in mature and stable condition. PDGF-b, adrenaline, Ang1, transforming growth factor B (TGF-b), and NOTCH lead the cells to be enveloped by pericytes. Tissue inhibitors of metalloproteins (TIMPs) and plasminogen inhibitor-1 activators (PAI-1) which inhibits protease lead to basement membrane deposition and reconnect to guarantee dynamic flow distribution. If the blood vessels fail to perfuse, they will degenerate.

2. BIOLOGICAL MOLECULES

2.1. The VEGF Family

Despite complicacy of angiogenesis and relating processes, it is worth noting that VEGF play a leading role in this process. The VEGF family consists limited members, whose non-conciseness makes it different from other angiogenesis super-families. VEGF, also defined as VEGF-a, is the key component, which regulate angiogenesis through signal transduction of VEGF receptor 2 signal both in conditions of health and disease [47]. Neuroteins like Nrp1 and Nrp2 serve as co-receptors of VEGF, which stimulates the activity of VEGFR-2 *via* an independent signal [48]. Equivalent to the absence of VEGFR-2, the lack of VEGF can also lead to the discontinuation of vascular development. VEGF gradient was established by both matrix-binding and soluble subtypes. Tip cells up-regulated the production of DLL4, thus activating the NOTCH signals in stalk cells. Thus lessen the expression of VEGFR-2, which weakened the response of stalk cells to VEGF, ensuring that tip cells occupied a dominant position [49]. Soluble VEGF subtypes promote vasodilation, while matrix-binding subtypes promote vascular to branch. Tumors, stromal cells and medullary cells releases paracrine VEGF, which stimulate branches and causes vascular abnormalities in tumors [50], while ECs release autocrine VEGF to balance vascular homeostasis. New studies suggest biological effects of VEGFR-2 signaling is based on its subcellular localization. For instance, to process arterial morphogenesis by VEGF, VEGFR-2 signal must be transformed from the intracellular region [51]. Activation of VEGFR-2 mutation can lead to hemangioma. Pathological angiogenesis occurs due to the genetic polymorphisms of VEGF. Blockage of VEGF signal can target angiogenesis in human malignant and eye diseases. Transform of VEGF gene increases vascular growth in ischemic tissue, but is often accompanied with abnormal leakage and vascular aberrance.

Combination of VEGF-C to VEGFR-2 and VEGFR-3 receptors activate vascular tip cells [52]. VEGFR-3 is indispensable in early embryonic angiogenesis, but later became main mediator of lymph-angiogenesis. In zebrafish, the first embryonic vein is produced by isolating vein ECs origin form an ordinary precursor vessel. The germination of vein ECs is confined by VEGFR-2, but stimulated by VEGFR-3 [53]. Vein-derived angiogenesis of arterial branch also requires VEGFR-3 signal. After applying Anti-VEGFR-3 antibodies, receptor dimerization or ligand binding was surpassed, thus decreased the growth of tumors. Enhanced inhibition of tumor growth was obtained by blocking VEGFR-2, suggesting another potential anti- angiogenesis site [52].

PLGF, a VEGF homologue was considered as an important angiogenic factor. However, PLGF is necessary for angiogenic development and is only connected with disease. PLGF is a multifunctional cytokine that promotes angiogenesis through direct and consequential mechanisms. PLGF activate medullary cells, stromal cells and bone marrow-derived endothelial progenitor cells, thus produced environment for activating cancer cells. Polarization of tumor-associated macrophages (TAMs) change cause PLGF to loss efficiency, thus improves vascular maturation, and promotes the react to chemotherapy. Usage of anti-PLGF antibody revealed the anti-angiogenesis effect of gene PLGF deficiency in mouse tumor model and intraocular neovascularization. However, in transplantable tumor model, different PLGF blocking protocols failed to repress the development of tumors. Therefore, the PLGF blockade, as therapeutic candidate for cancer patients needs to be further discussed. Preclinical research proved, both transfer of PLGF protein or gene promotes the revascularization of ischemic tissue.

The absence of VEGF-B, a member of the VEGF family in mice, does not affect angiogenesis during growth and cannot make up for the blocking effect of VEGF after birth. VEGF-B restricts angiogenesis only in specific tissues for instance the heart, but it stimulates neuronal system survival and balance metabolic dynamics. VEGF-B has different effects on pathological angiogenesis, it can increase the blood vessels development without causing side effects like more permeability and leakage.

The participant of VEGFR-1 receptor (FLT-1) in angiogenesis is unknown [54]. VEGFR-1 behave in the form of membrane anchored signal transduction and soluble secretion (sFLT-1). Once combined, sFLT-1 serve as guidance for new branches or completely inhibit germination. VEGFR-1 can be considered a bait for VEGF due to its minimal tyrosine kinase activity, which regulate soluble VEGF that can be used to provoke VEGFR-2, and interpret how lack of VEGFR-1 leading to vascular overgrowth. However, VEGFR-1 signals in ECs, myelocytes and stromal cells lead to pathological angiogenesis [54].

2.2. PDGF Family

Blood vessels must mature and enveloped in mural cells to develop function. Growth factors including PDGFs, TGF-β, and angiopoietins are detected in this process. Then, endothelial cell channels are stabilized by process of angiogenic ECs releasing PDGF-B to chemically attract PDGF receptor β (PDGFR-β)+ pericytes [55, 56]. Therefore, lack of pericytes after PDGF-B inhibition leads to distortion, vascular leakage, bleeding and microaneurysm formation. PDGF-B gene knockout in mice can lead to the fragility and excessive expansion of tumor blood vessels, and PDGFR-β subtype mice have deficient perivascular cells surrounding the cerebrovascular, resulting in blood-brain barrier (BBB) defects and neurodegenerative damage due to the leakage of toxic substances. Tumor-derived PDGF-B recruits pericytes indirectly by up-regulating stromal cell-derived factor-1α (SDF-1α; CXCL12 coding). In addition to local sources, pericytes originate from PDGFR-β+ pericyte progenitor cells around blood vessels and recruit from marrow. Blockade of PDGFR-β signal in mural cells, VEGF decreases presence of pericyte coverage, and abnormalities of tumor blood vessels.

Inhibition of PDGFR reduces the growth of tumors by causing pericytes detachment, leading to the degeneration of immature blood vessels [57]. Other pericyte-deficient murine strains lacking proteoglycan NG2 (CSPG4) produce smaller tumor vessels and variant tumors. However, PDGF-B overexpression in mice slowed down growth of tumor by stimulating pericyte accumulation as while as promoting endothelial cell growth. The survival of ECs depends on the amount of VEGF in pericytes, pericytes prevents ECs to discard VEGF and resist the blockade of VEGF, which depends on intimate pericyte-endothelial interaction, because only when VEGF is produced from pericytes instead of distant cancer cells, can inhibition of PDGF-B decrease pericyte warping and blood vessels growing. Application of a multi-target receptor tyrosine kinase inhibitor (TKIs) have shown that inhibition of PDGF-B increases sensitivity of mature blood vessels to blockade of VEGF by depleting pericytes. Further researches applying of specific inhibitors revealed that multi-target therapy is less effective than sole anti-VEGF treatment [58].

2.3. Angiopoietin

Angiopoietin plays an important role in the regulating stability of vascular network. Two different angiopoietins exist and displayed antagonistic effects on angiogenesis. Angiopoietin 1 (Ang1) is produced by non-vascular normal cells, vascular mural cells and tumor cells [59], which is helpful to maintain the stable state of blood vessels [60]. Ang1 mediates a series of effects during its receptor activation, such as strengthen of endothelial cell junction, stimulation of endothelial cell survival and promoting endothelial-mural cell reaction [61]. Angiopoietin 2 (Ang2) is dominantly produced by ECs during vascular remodeling and competes with Ang1 for the same receptor as binding ligands. The influence of Ang2 stimulation is decided on the existence of VEGF. Ang2 reaction in the presence of VEGF sensitizes ECs to other angiogenic factors-mediated proliferation signals [61], increases basal layer remodeling and stimulates endothelial cell migration, which leads to vascular remodeling. Ang2 can cause endothelial cell death and vascular degeneration in the absence of VEGF.

2.4. The bFGF and The FGF Superfamily

Basic fibroblast growth factor (bFGF) is a multiple function cytokine which give play in the progress of angiogenesis by stimulating the differentiation, migration and survival of ECs. Besides promoting capillary growth, bFGF induces proliferation of smooth muscle cells thus stimulate the growth of larger vessels. The FGFs superfamily along with their receptors participated in various biological functions. The bFGF is one of the earliest detected angiogenesis factors, which has the characteristics of angiogenesis and arterio-genesis. FGF9 stimulates angiogenesis in bone repair by stimulating other types of cells to release angiogenic factors [62]. Low levels of FGF are needed to maintain vascular completion. Blockade of FGFR signal of immobile ECs leads to vascular abruption [63]. Variant FGF signal stimulate angiogenesis and assist tumor vessels to withdraw from inhibitors of epidermal growth factor receptor (EGFR) and VEGF. The study of specific FGF or FGFR inhibitors to block angiogenesis is fall behind, which may be due to the large amount of membership in FGF superfamily and FGF-1 or FGF-2 deficient mice fail to abrupt vascular development. FGF gene therapy has been applied to treat angiogenesis, however failed to sustain in clinical success.

2.5. Integrins and Proteases

Extracellular matrix (ECM) provides a physical structure which connect vascular cells and extracellular tissues. ECs have the mechanism of interacting with matrix and changing matrix. Integrin is a heteromeric two receptor that mediates adhesion molecules ECM and immunoglobulin superfamily [64]. Integrin regulates angiogenesis through other mechanisms besides inducing signals by connecting ECM elements. Since integrin can bind to various surrounding molecules and deliver signals in a two-way, integrin acts as a "center" to coordinate the behaviour of vascular smooth muscle cells in angiogenesis. Therefore, the interaction of integrin with VEGF, Ang1, FGFs and their receptors such as VEGFR-2 and FGFRs promotes vascular development. Integrin can up-regulate protease activity of invasive terminal cells, and stimulates vascular to mature by modulating the interplay with pericytes, ECs and the basement membrane.

ECs share the same base membrane with pericytes cells, which physically restrain these cells and puts them into a stationed condition due to the anti-proliferation characteristic of ECM components. In the branching process, extracellular matrix hydrolysis enzymes protein remodel and release these cells, making them move unrestricted, and transforming the properties of membrane base into a vasculogennesis atmosphere. Matrix metalloproteinase (MMP) family regulate angiogenesis *via* several mechanisms. They stimulate migration of endothelial cell by reconstructing membrane base by protein hydrolysis, or direct hydrolysis of matrix protein such as membrane type 1-MMP, or expose hidden motif sites through chemotactic in ECM [65]. MMPs and fibrinolytic enzymes can also stimulate vasculogennesis by revealing VEGF and FGF from matrix pool. MMPs degrade subtype of VEGF, thus increase soluble VEGF circulation and enlarges the blood vessels, while the matrix-bound VEGF of MMP resistance supports the branch. Macrophages, mast cells and neutrophils activate VEGF mediated *via* MMP9 to initiate angiogenesis. Proteinases like MMP9 also take part in the regulation of bone marrow progenitor cells by increasing soluble kit ligands termed as stalk cell factors or SCF [66]. In view of the potentially destructive nature of protease, its activity should be strictly commanded. The loss of plasminogen activator inhibitor-1(PAI-1) impedes vascular to branch, because redundant ECM decomposition needs matrix supply for sproutation. Besides, membrane base requires activity of MMP inhibitors for instance TIMPs to deposit during vascular development. Degradation of ECM proportions can increase anti-angiogenesis components, the biological effects of protease inhibitors must be carefully evaluated.

2.6. Junctional Molecules

Communication between cell and cell is the basis of vascular synchronization. This coordination is achieved through gap junctions intercellular communication (GJIC), which are assembled by connexins. Connexins signal through the perfusion status of downstream tissues of upstream blood supply vessels to holdback shunting. Beside indirectly connexin communications, pericytes and ECs also process directly communication sites. Static ECs are a collection of intercellular connected cells in the form of layer, while angiogenetic ECs managed to scatter their binding and promote migrate. VE-cadherin deficiency does not impede the maturation of blood vessels, instead it causes abnormal vascular remodeling or integrity. Combination of VE-cadherin with CD34 in cells is essential to form lumen [67]. In stationary phalangeal VE-cadherin stimulates TGFR signals by impeding VEGFR-2 pathways and promotes vascular stability. It is worth noting that the oxygen sensor controls the VE-cadherin production in the feedback circuit, thus optimize the perfusion blood vessels in oxygen deficiency circumstances [68]. Cadherin functions to rigid the connection of pericytes cells and ECs. In the period of germination, VEGF along with angiogenesis factors impeded the adhesion function of VE-cadherin between adjacent cells. Meanwhile VE-cadherin orientation in the filament functions to stimulate connections between cells from the bud and apical cells. New epitopes sites of VE-cadherin were recognized by antibodies, while exposer of VE-cadherin to adhesion junctions during germination provide an opportunity for inhibition of endothelial cell development and maintain stability of ECs.

3. THE MARKERS AND TESTING OF ANGIOGENESIS

Angiogenesis develops on the basis of the original microvascular bed, mainly capillaries and venules. Angiogenesis is regulated and influenced by a variety of factors, the acidic environment and high lactic acid concentration caused by hypoxia and low perfusion may be the most important factors. They induce VEGF, PDGF, TGF-1, and other secretions of VEGF. Heparin, thromboxane and proteoglycan cleavage products from the extracellular matrix can cause the release of FGF, PDGF and TGF, and induce the formation of new blood vessels [69 - 72].

As mentioned above, the angiogenesis process can be divided into five stages: 1. vascular secretion of protease, basement membrane collapse; 2. migration of ECs, through the basement membrane, into the perivascular matrix; 3. endothelial cell proliferation, interconnection; 4. proliferation The ECs form a three-dimensional structure of the lumen; 5. the basement membrane forms a lumen surrounding the lumen and conforms to each other to form a tube and a mesh. It should be noted that angiogenesis occurs in the matrix surrounding the blood vessels, and the necessary understanding of the extracellular matrix should be made. Endothelial migration, protein degradation and growth are three important factors in angiogenesis. The common feature of angiogenesis-inducing agents such as bFGF is that it has specific stimulatory influence on migration of ECs, protein degradation and growth. The movement of ECs causes chemotaxis to vascular stimulating factors and aligns to form vascular sprouts. Proteolysis allows vascular sprouts to travel and expand within the cell matrix. The proliferative activity of ECs ensures the formation of the vascular lumen and establishes a new blood circulation pathway.

Therefore, collagen in the cell matrix, enzymes involved in protein degradation, and various promoting factors can become potential markers in angiogenesis, indicating changes in the vascularization process.

3.1. Markers in Extracellular Matrix

3.1.1. Fibronectin

Angiogenesis FN is the most characteristic non-collagen support molecule. FN is a precursor of endothelial cell migration. In the early stage of angiogenesis, the content of intravascular FN (fibronectin fibronectin) increased, and the FN of extracellular matrix showed a strong staining reaction. FN could promote the migration of this endothelial cell. It appears to be the most important component of the matrix during germination of blood vessels, but its content rapidly decreases as the blood vessels mature. Researches revealed that microvascular ECs grow faster in FN-containing medium than other stromal media [73, 74]. In the current study, the detection methods of fibronectin content were Western Blot, enzyme-linked immunosorbent assay and immunohistochemical staining [75 - 77]. Biopsy specimens were immunohistochemically stained to detect levels of fibronectin. The focus of ELISA is to allow the antibody to bind to the enzyme complex and then detect it by colour development. It can be used to detect the content of fibronectin in serum, plasma and related liquid samples.

3.1.2. Polysaccharide

Feinberg *et al.* reported that when the chicken embryo limb germinated, there was a large avascular zone around it, which contained a lot of hyaluronic acid. A high level of hyaluronic acid in the graft can also produce avascular bands. In other cases of tumor growth, metastasis, chronic inflammation, *etc.*, hyaluronic acid can act as a direct or indirect angiogenesis inhibitor. Heparin stimulates the migration and angiogenesis of capillary ECs. Heparin binds strongly to ECs, and each endothelial cell can bind 106 heparin molecules. In the early stage, radioimmunoassay, which uses isotopically labeled antigens to compete with antibodies, can be used to study the reaction of the body to antigenic substances and to detect hyaluronic acid. The ELISA method is used to detect the content of hyaluronic acid in serum.

3.1.3. Promoting Factor

In recent years, basic research on angiogenesis began with angiogenesis-promoting factors, such as the discovery of VEGF and its receptors (ECs only), which promoted basic research on angiogenesis and formed a climax to study angiogenesis. Many factors participate in maturation of blood vessels. The main ones are tumor angiogenesis (TAF) such as VEGF, PDGF, bFGF and TGF, heparin binding endothelial cell growth factors, and tumor necrosis factor (TNF).

The common feature of angiogenesis-inducing agents such as bFGF is the stimulatory effects on protein degradation, growth and migration of ECs. The movement of ECs causes chemotaxis to vascular stimulating factors and aligns to form vascular sprouts. Proteolysis allows vascular sprouts to travel and expand within the cell matrix. The proliferation of ECs ensures development of blood vessels, thus establishes a new blood circulation pathway. The levels of promoting factors in serum are usually determined using an enzyme-linked immunosorbent assay. VEGF isoforms are detected through alternative RNA splicing, the major VEGF isoforms detected in human body is VEGF165 and VEGF121. However, functions of different VEGF isoforms remain further study to detect whether VEGF isoforms are possible markers for anti-angiogenic therapy.

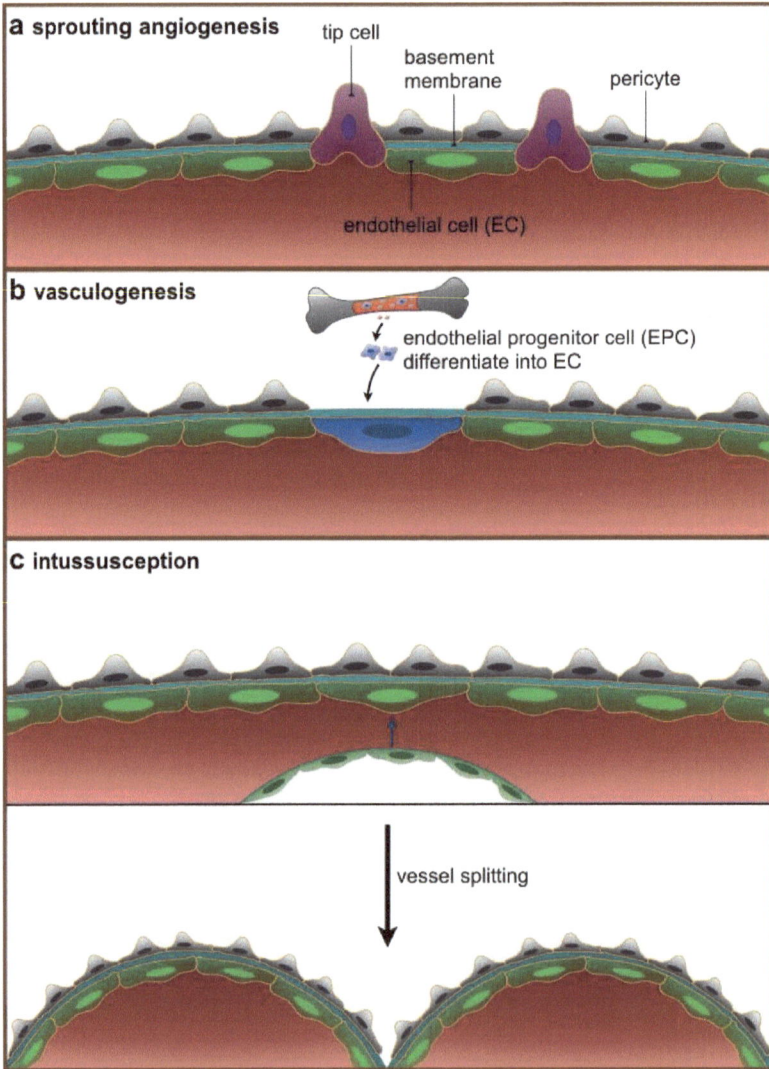

Fig. (1). Angiogenesis refers to the formation of new blood vessels from the existing vascular network. These blood vessels can be formed by sprouting from the pre-existing blood vessels, or they can be formed by intussusception, which means pillar-shaped tissue inserted into the existing capillary to divide the blood vessels. **a**, new vessels occur by sprouting angiogenesis. **b**, by recruiting endothelial progenitor cells (EPCs) from bone marrow which finally differentiate into endothelial cells (ECs), new vessels occur. **c**. an existing vessel split into two vessels, which is called intussusception.

3.1.4. Related Proteases

The main condition for angiogenesis and production is the degradation of matrix components, so that ECs can migrate and germinate. The related proteases mainly include three major categories: 1 serine proteases (plasminogen activator, leukocyte elastase, cathepsin G, *etc.*); 2 metalloproteinases Classes (interstitial collagenase, type IV collagenase, interstitial lysin, *etc.*); 3 cysteine proteases (endogenous and exogenous peptidases) to numerous structural molecules of the membrane base for instance fibronectin, various types of collagen, laminin and core proteins of protein mucopolysaccharides all have degradation-destructive effects, which help to clean and prepare conditions for angiogenesis. Plasma enzyme immunoassay was used to observe the expression of related proteases in plasma during angiogenesis, or immunohistochemical staining was used to detect related proteases in tissues.

3.1.5. Specific Substance Produced by Angiogenesis

In the process of micro-angiogenesis, not only the growth and secretion of angiogenic factors, but also endogenous inhibition of angiogenic factors, such as thrombospondin (TSP), metalloproteinase inhibitors, platelet factor 4, *etc.* are involved in the process. In this process, not only various changes occur in vascular ECs, but also many changes occur in the extracellular matrix. Along with the angiogenesis, certain substances have been produced, such as collagen, Decorin, SPARC, cell moving protein (AAMP), protein solubilizing enzyme, *etc.*

3.1.5.1. Collagen 1

Collagen 1 is not expressed in normal tissue vessels, and the normal extracapillary basement membrane is composed of collagen 4 and laminin. When vascular ECs begin to proliferate, it secretes collagen 1 and, therefore, collagen 1 is a blood vessel. Specific product when budding grows.

3.1.5.2. Decorin

Decorin is a 100 kD molecular weight chondroitin sulfated sugar that is synthesized and secreted while vascular ECs synthesize collagen 1. Normal ECs do not express, and it synergizes with collagen 1 to cause sprouting and lumen formation of vascular ECs.

3.1.5.3. Sparc

SPARC is a protein with a molecular weight of 43 kD, which is secreted protein (acid and rich in cysteine). After secretion by protein 1 and decirun, after neovascularization, synthesis and secretion are formed when the lumen is formed.

It is involved in the rearrangement and assembly of new ECs.

CONSENT FOR PUBLICATION

Not applicable.

CONFLICT OF INTEREST

The authors confirm that the contents of this chapter have no conflict of interest.

ACKNOWLEDGEMENTS

This study was supported by National Key R&D Program of China (2019YFA0110600) and National Natural Science Foundation of China (81970986, 81771125).

REFERENCES

[1]   Simons M. Angiogenesis: where do we stand now? Circulation 2005; 111(12): 1556-66.
        [http://dx.doi.org/10.1161/01.CIR.0000159345.00591.8F] [PMID: 15795364]

[2]   Carmeliet P, Jain RK. Molecular mechanisms and clinical applications of angiogenesis. Nature 2011; 473(7347): 298-307.
        [http://dx.doi.org/10.1038/nature10144] [PMID: 21593862]

[3]   Takahashi T, Kalka C, Masuda H, *et al.* Ischemia- and cytokine-induced mobilization of bone marrow-derived endothelial progenitor cells for neovascularization. Nat Med 1999; 5(4): 434-8.
        [http://dx.doi.org/10.1038/7434] [PMID: 10202935]

[4]   Tepper OM, Capla JM, Galiano RD, *et al.* Adult vasculogenesis occurs through *in situ* recruitment, proliferation, and tubulization of circulating bone marrow-derived cells. Blood 2005; 105(3): 1068-77.
        [http://dx.doi.org/10.1182/blood-2004-03-1051] [PMID: 15388583]

[5]   Risau W. Mechanisms of angiogenesis. Nature 1997; 386(6626): 671-4.
        [http://dx.doi.org/10.1038/386671a0] [PMID: 9109485]

[6]   Sabin FR. Preliminary note on the differentiation of angioblasts and the method by which they produce blood-vessels, blood-plasma and red blood-cells as seen in the living chick. 1917. J Hematother Stem Cell Res 2002; 11(1): 5-7.
        [http://dx.doi.org/10.1089/152581602753448496] [PMID: 11846999]

[7]   Amaya E. The hemangioblast: a state of competence. Blood 2013; 122(24): 3853-4.
        [http://dx.doi.org/10.1182/blood-2013-10-533075] [PMID: 24311714]

[8]   Huber TL, Kouskoff V, Fehling HJ, Palis J, Keller G. Haemangioblast commitment is initiated in the primitive streak of the mouse embryo. Nature 2004; 432(7017): 625-30.
        [http://dx.doi.org/10.1038/nature03122] [PMID: 15577911]

[9]   Lacaud G, Keller G, Kouskoff V. Tracking mesoderm formation and specification to the hemangioblast *in vitro.* Trends Cardiovasc Med 2004; 14(8): 314-7.
        [http://dx.doi.org/10.1016/j.tcm.2004.09.004] [PMID: 15596108]

[10]   Kennedy M, Firpo M, Choi K, *et al.* A common precursor for primitive erythropoiesis and definitive haematopoiesis. Nature 1997; 386(6624): 488-93.
        [http://dx.doi.org/10.1038/386488a0] [PMID: 9087406]

[11]   Choi K, Kennedy M, Kazarov A, Papadimitriou JC, Keller G. A common precursor for hematopoietic

and endothelial cells. Development 1998; 125(4): 725-32.
[PMID: 9435292]

[12] Kurosawa H. Methods for inducing embryoid body formation: *in vitro* differentiation system of embryonic stem cells. J Biosci Bioeng 2007; 103(5): 389-98.
[http://dx.doi.org/10.1263/jbb.103.389] [PMID: 17609152]

[13] Liu F, Kang I, Park C, *et al.* ER71 specifies Flk-1+ hemangiogenic mesoderm by inhibiting cardiac mesoderm and Wnt signaling. Blood 2012; 119(14): 3295-305.
[http://dx.doi.org/10.1182/blood-2012-01-403766] [PMID: 22343916]

[14] Liu F, Li D, Yu YY, *et al.* Induction of hematopoietic and endothelial cell program orchestrated by ETS transcription factor ER71/ETV2. EMBO Rep 2015; 16(5): 654-69.
[http://dx.doi.org/10.15252/embr.201439939] [PMID: 25802403]

[15] Watabe T, Miyazono K. Roles of TGF-beta family signaling in stem cell renewal and differentiation. Cell Res 2009; 19(1): 103-15.
[http://dx.doi.org/10.1038/cr.2008.323] [PMID: 19114993]

[16] Park KS, Cha Y, Kim CH, *et al.* Transcription elongation factor Tcea3 regulates the pluripotent differentiation potential of mouse embryonic stem cells *via* the Lefty1-Nodal-Smad2 pathway. Stem Cells 2013; 31(2): 282-92.
[http://dx.doi.org/10.1002/stem.1284] [PMID: 23169579]

[17] Lee D, Park C, Lee H, *et al.* ER71 acts downstream of BMP, Notch, and Wnt signaling in blood and vessel progenitor specification. Cell Stem Cell 2008; 2(5): 497-507.
[http://dx.doi.org/10.1016/j.stem.2008.03.008] [PMID: 18462699]

[18] Marcelo KL, Goldie LC, Hirschi KK. Regulation of endothelial cell differentiation and specification. Circ Res 2013; 112(9): 1272-87.
[http://dx.doi.org/10.1161/CIRCRESAHA.113.300506] [PMID: 23620236]

[19] Cha Y, Heo SH, Ahn HJ, *et al.* Tcea3 regulates the vascular differentiation potential of mouse embryonic stem cells. Gene Expr 2013; 16(1): 25-30.
[http://dx.doi.org/10.3727/105221613X13776146743343] [PMID: 24397209]

[20] Drake CJ, Brandt SJ, Trusk TC, Little CD. TAL1/SCL is expressed in endothelial progenitor cells/angioblasts and defines a dorsal-to-ventral gradient of vasculogenesis. Dev Biol 1997; 192(1): 17-30.
[http://dx.doi.org/10.1006/dbio.1997.8751] [PMID: 9405094]

[21] Drake CJ, Fleming PA. Vasculogenesis in the day 6.5 to 9.5 mouse embryo. Blood 2000; 95(5): 1671-9.
[http://dx.doi.org/10.1182/blood.V95.5.1671.005k39_1671_1679] [PMID: 10688823]

[22] Ferguson JE III, Kelley RW, Patterson C. Mechanisms of endothelial differentiation in embryonic vasculogenesis. Arterioscler Thromb Vasc Biol 2005; 25(11): 2246-54.
[http://dx.doi.org/10.1161/01.ATV.0000183609.55154.44] [PMID: 16123328]

[23] Sato Y. Dorsal aorta formation: separate origins, lateral-to-medial migration, and remodeling. Dev Growth Differ 2013; 55(1): 113-29.
[http://dx.doi.org/10.1111/dgd.12010] [PMID: 23294360]

[24] Quillien A, Moore JC, Shin M, *et al.* Distinct Notch signaling outputs pattern the developing arterial system. Development 2014; 141(7): 1544-52.
[http://dx.doi.org/10.1242/dev.099986] [PMID: 24598161]

[25] Iozzo RV. Basement membrane proteoglycans: from cellar to ceiling. Nat Rev Mol Cell Biol 2005; 6(8): 646-56.
[http://dx.doi.org/10.1038/nrm1702] [PMID: 16064139]

[26] Miner JH. Laminins and their roles in mammals. Microsc Res Tech 2008; 71(5): 349-56.
[http://dx.doi.org/10.1002/jemt.20563] [PMID: 18219670]

[27]   Asahara T, Murohara T, Sullivan A, *et al.* Isolation of putative progenitor endothelial cells for angiogenesis. Science 1997; 275(5302): 964-7.
[http://dx.doi.org/10.1126/science.275.5302.964] [PMID: 9020076]

[28]   Asahara T, Kawamoto A. Endothelial progenitor cells for postnatal vasculogenesis. Am J Physiol Cell Physiol 2004; 287(3): C572-9.
[http://dx.doi.org/10.1152/ajpcell.00330.2003] [PMID: 15308462]

[29]   Pober JS. Just the FACS or stalking the elusive circulating endothelial progenitor cell. Arterioscler Thromb Vasc Biol 2012; 32(4): 837-8.
[http://dx.doi.org/10.1161/ATVBAHA.112.246280] [PMID: 22423031]

[30]   Rehman J, Li J, Orschell CM, March KL. Peripheral blood "endothelial progenitor cells" are derived from monocyte/macrophages and secrete angiogenic growth factors. Circulation 2003; 107(8): 1164-9.
[http://dx.doi.org/10.1161/01.CIR.0000058702.69484.A0] [PMID: 12615796]

[31]   Hirschi KK, Ingram DA, Yoder MC. Assessing identity, phenotype, and fate of endothelial progenitor cells. Arterioscler Thromb Vasc Biol 2008; 28(9): 1584-95.
[http://dx.doi.org/10.1161/ATVBAHA.107.155960] [PMID: 18669889]

[32]   Recchioni R, Marcheselli F, Antonicelli R, *et al.* Physical activity and progenitor cell-mediated endothelial repair in chronic heart failure: Is there a role for epigenetics? Mech Ageing Dev 2016; 159: 71-80.
[http://dx.doi.org/10.1016/j.mad.2016.03.008] [PMID: 27015708]

[33]   Rajasekar P, O'Neill CL, Eeles L, Stitt AW, Medina RJ. Epigenetic changes in endothelial progenitors as a possible cellular basis for glycemic memory in diabetic vascular complications. J Diabetes Res 2015; 2015: 436879.
[http://dx.doi.org/10.1155/2015/436879] [PMID: 26106624]

[34]   Hill JM, Zalos G, Halcox JP, *et al.* Circulating endothelial progenitor cells, vascular function, and cardiovascular risk. N Engl J Med 2003; 348(7): 593-600.
[http://dx.doi.org/10.1056/NEJMoa022287] [PMID: 12584367]

[35]   Yoder MC, Mead LE, Prater D, *et al.* Redefining endothelial progenitor cells *via* clonal analysis and hematopoietic stem/progenitor cell principals. Blood 2007; 109(5): 1801-9.
[http://dx.doi.org/10.1182/blood-2006-08-043471] [PMID: 17053059]

[36]   Tanaka R, Wada M, Kwon SM, *et al.* The effects of flap ischemia on normal and diabetic progenitor cell function. Plast Reconstr Surg 2008; 121(6): 1929-42.
[http://dx.doi.org/10.1097/PRS.0b013e3181715218] [PMID: 18520878]

[37]   Yoder MC. Human endothelial progenitor cells. Cold Spring Harb Perspect Med 2012; 2(7): a006692.
[http://dx.doi.org/10.1101/cshperspect.a006692] [PMID: 22762017]

[38]   Yoder MC. Editorial: Early and late endothelial progenitor cells are miR-tually exclusive. J Leukoc Biol 2013; 93(5): 639-41.
[http://dx.doi.org/10.1189/jlb.0113004] [PMID: 23633478]

[39]   Basile DP, Yoder MC. Circulating and tissue resident endothelial progenitor cells. J Cell Physiol 2014; 229(1): 10-6.
[PMID: 23794280]

[40]   Majesky MW, Dong XR, Hoglund V, Daum G, Mahoney WM Jr. The adventitia: a progenitor cell niche for the vessel wall. Cells Tissues Organs (Print) 2012; 195(1-2): 73-81.
[http://dx.doi.org/10.1159/000331413] [PMID: 22005572]

[41]   Balaji S, King A, Crombleholme TM, Keswani SG. The role of endothelial progenitor cells in postnatal vasculogenesis: Implications for therapeutic neovascularization and wound healing. Adv Wound Care (New Rochelle) 2013; 2(6): 283-95.
[http://dx.doi.org/10.1089/wound.2012.0398] [PMID: 24527350]

[42] Blum A. Endothelial progenitor cells are affected by medications and estrogen. Isr Med Assoc J 2015; 17(9): 578-80.
[PMID: 26625551]

[43] Urbich C, Dimmeler S. Endothelial progenitor cells: characterization and role in vascular biology. Circ Res 2004; 95(4): 343-53.
[http://dx.doi.org/10.1161/01.RES.0000137877.89448.78] [PMID: 15321944]

[44] Wang R, Chadalavada K, Wilshire J, *et al.* Glioblastoma stem-like cells give rise to tumour endothelium. Nature 2010; 468(7325): 829-33.
[http://dx.doi.org/10.1038/nature09624] [PMID: 21102433]

[45] Swift MR, Weinstein BM. Arterial-venous specification during development. Circ Res 2009; 104(5): 576-88.
[http://dx.doi.org/10.1161/CIRCRESAHA.108.188805] [PMID: 19286613]

[46] Schaper W. Collateral circulation: past and present. Basic Res Cardiol 2009; 104(1): 5-21.
[http://dx.doi.org/10.1007/s00395-008-0760-x] [PMID: 19101749]

[47] Ferrara N. VEGF-A: a critical regulator of blood vessel growth. Eur Cytokine Netw 2009; 20(4): 158-63.
[http://dx.doi.org/10.1684/ecn.2009.0170] [PMID: 20167554]

[48] Neufeld G, Kessler O. The semaphorins: versatile regulators of tumour progression and tumour angiogenesis. Nat Rev Cancer 2008; 8(8): 632-45.
[http://dx.doi.org/10.1038/nrc2404] [PMID: 18580951]

[49] Phng LK, Gerhardt H. Angiogenesis: a team effort coordinated by notch. Dev Cell 2009; 16(2): 196-208.
[http://dx.doi.org/10.1016/j.devcel.2009.01.015] [PMID: 19217422]

[50] Stockmann C, Doedens A, Weidemann A, *et al.* Deletion of vascular endothelial growth factor in myeloid cells accelerates tumorigenesis. Nature 2008; 456(7223): 814-8.
[http://dx.doi.org/10.1038/nature07445] [PMID: 18997773]

[51] Lanahan AA, Hermans K, Claes F, *et al.* VEGF receptor 2 endocytic trafficking regulates arterial morphogenesis. Dev Cell 2010; 18(5): 713-24.
[http://dx.doi.org/10.1016/j.devcel.2010.02.016] [PMID: 20434959]

[52] Tvorogov D, Anisimov A, Zheng W, *et al.* Effective suppression of vascular network formation by combination of antibodies blocking VEGFR ligand binding and receptor dimerization. Cancer Cell 2010; 18(6): 630-40.
[http://dx.doi.org/10.1016/j.ccr.2010.11.001] [PMID: 21130043]

[53] Herbert SP, Huisken J, Kim TN, *et al.* Arterial-venous segregation by selective cell sprouting: an alternative mode of blood vessel formation. Science 2009; 326(5950): 294-8.
[http://dx.doi.org/10.1126/science.1178577] [PMID: 19815777]

[54] Schwartz JD, Rowinsky EK, Youssoufian H, Pytowski B, Wu Y. Vascular endothelial growth factor receptor-1 in human cancer: concise review and rationale for development of IMC-18F1 (Human antibody targeting vascular endothelial growth factor receptor-1). Cancer 2010; 116(4) (Suppl.): 1027-32.
[http://dx.doi.org/10.1002/cncr.24789] [PMID: 20127948]

[55] Hellberg C, Ostman A, Heldin CH. PDGF and vessel maturation. Recent Results Cancer Res 2010; 180: 103-14.
[http://dx.doi.org/10.1007/978-3-540-78281-0_7] [PMID: 20033380]

[56] Gaengel K, Genové G, Armulik A, Betsholtz C. Endothelial-mural cell signaling in vascular development and angiogenesis. Arterioscler Thromb Vasc Biol 2009; 29(5): 630-8.
[http://dx.doi.org/10.1161/ATVBAHA.107.161521] [PMID: 19164813]

[57]   Bergers G, Song S, Meyer-Morse N, Bergsland E, Hanahan D. Benefits of targeting both pericytes and endothelial cells in the tumor vasculature with kinase inhibitors. J Clin Invest 2003; 111(9): 1287-95.
[http://dx.doi.org/10.1172/JCI200317929] [PMID: 12727920]

[58]   Nisancioglu MH, Betsholtz C, Genové G. The absence of pericytes does not increase the sensitivity of tumor vasculature to vascular endothelial growth factor-A blockade. Cancer Res 2010; 70(12): 5109-15.
[http://dx.doi.org/10.1158/0008-5472.CAN-09-4245] [PMID: 20501841]

[59]   Danza K, Pilato B, Lacalamita R, *et al.* Angiogenetic axis angiopoietins/Tie2 and VEGF in familial breast cancer. Eur J Hum Genet 2013; 21(8): 824-30.
[http://dx.doi.org/10.1038/ejhg.2012.273] [PMID: 23232696]

[60]   Folkman J. Angiogenesis: an organizing principle for drug discovery? Nat Rev Drug Discov 2007; 6(4): 273-86.
[http://dx.doi.org/10.1038/nrd2115] [PMID: 17396134]

[61]   Sakurai T, Kudo M. Signaling pathways governing tumor angiogenesis. Oncology 2011; 81 (Suppl. 1): 24-9.
[http://dx.doi.org/10.1159/000333256] [PMID: 22212932]

[62]   Beenken A, Mohammadi M. The FGF family: biology, pathophysiology and therapy. Nat Rev Drug Discov 2009; 8(3): 235-53.
[http://dx.doi.org/10.1038/nrd2792] [PMID: 19247306]

[63]   Murakami M, Nguyen LT, Zhuang ZW, *et al.* The FGF system has a key role in regulating vascular integrity. J Clin Invest 2008; 118(10): 3355-66.
[http://dx.doi.org/10.1172/JCI35298] [PMID: 18776942]

[64]   Desgrosellier JS, Cheresh DA. Integrins in cancer: biological implications and therapeutic opportunities. Nat Rev Cancer 2010; 10(1): 9-22.
[http://dx.doi.org/10.1038/nrc2748] [PMID: 20029421]

[65]   Bergers G, Brekken R, McMahon G, *et al.* Matrix metalloproteinase-9 triggers the angiogenic switch during carcinogenesis. Nat Cell Biol 2000; 2(10): 737-44.
[http://dx.doi.org/10.1038/35036374] [PMID: 11025665]

[66]   Heissig B, Hattori K, Dias S, *et al.* Recruitment of stem and progenitor cells from the bone marrow niche requires MMP-9 mediated release of kit-ligand. Cell 2002; 109(5): 625-37.
[http://dx.doi.org/10.1016/S0092-8674(02)00754-7] [PMID: 12062105]

[67]   Strilić B, Kucera T, Eglinger J, *et al.* The molecular basis of vascular lumen formation in the developing mouse aorta. Dev Cell 2009; 17(4): 505-15.
[http://dx.doi.org/10.1016/j.devcel.2009.08.011] [PMID: 19853564]

[68]   Mazzone M, Dettori D, de Oliveira RL, *et al.* Heterozygous deficiency of PHD2 restores tumor oxygenation and inhibits metastasis *via* endothelial normalization. Cell 2009; 136(5): 839-51.
[http://dx.doi.org/10.1016/j.cell.2009.01.020] [PMID: 19217150]

[69]   Djonov V, Schmid M, Tschanz SA, Burri PH. Intussusceptive angiogenesis: its role in embryonic vascular network formation. Circ Res 2000; 86(3): 286-92.
[http://dx.doi.org/10.1161/01.RES.86.3.286] [PMID: 10679480]

[70]   Stephan CC, Brock TA. Vascular endothelial growth factor, a multifunctional polypeptide. P R Health Sci J 1996; 15(3): 169-78.
[PMID: 8994281]

[71]   Ijichi A, Sakuma S, Tofilon PJ. Hypoxia-induced vascular endothelial growth factor expression in normal rat astrocyte cultures. Glia 1995; 14(2): 87-93.
[http://dx.doi.org/10.1002/glia.440140203] [PMID: 7558244]

[72]   Beckner ME, Krutzsch HC, Stracke ML, Williams ST, Gallardo JA, Liotta LA. Identification of a new

immunoglobulin superfamily protein expressed in blood vessels with a heparin-binding consensus sequence. Cancer Res 1995; 55(10): 2140-9.
[PMID: 7743515]

[73] Cooper TP, Sefton MV. Fibronectin coating of collagen modules increases *in vivo* HUVEC survival and vessel formation in SCID mice. Acta Biomater 2011; 7(3): 1072-83.
[http://dx.doi.org/10.1016/j.actbio.2010.11.008] [PMID: 21059413]

[74] Giroux S, Tremblay M, Bernard D, *et al.* Embryonic death of Mek1-deficient mice reveals a role for this kinase in angiogenesis in the labyrinthine region of the placenta. Curr Biol 1999; 9(7): 369-72.
[http://dx.doi.org/10.1016/S0960-9822(99)80164-X] [PMID: 10209122]

[75] Monti M, Iommelli F, De Rosa V, *et al.* Integrin-dependent cell adhesion to neutrophil extracellular traps through engagement of fibronectin in neutrophil-like cells. PLoS One 2017; 12(2): e0171362.
[http://dx.doi.org/10.1371/journal.pone.0171362] [PMID: 28166238]

[76] Oegema TR Jr, Johnson SL, Aguiar DJ, Ogilvie JW. Fibronectin and its fragments increase with degeneration in the human intervertebral disc. Spine 2000; 25(21): 2742-7.
[http://dx.doi.org/10.1097/00007632-200011010-00005] [PMID: 11064518]

[77] Whitacre CM. Application of Western blotting to the identification of metallothionein binding proteins. Anal Biochem 1996; 234(1): 99-102.
[http://dx.doi.org/10.1006/abio.1996.0056] [PMID: 8742089]

CHAPTER 2

Effects of Microenvironment Factors on Angiogenesis

Changyue Xue and **Xiaoxiao Cai**[*]

State Key Laboratory of Oral Diseases, West China Hospital of Stomatology, Sichuan University, Chengdu 610041, China

Abstract: Angiogenesis is a vital step for complete organ engineering, and the microenvironment is one of the four basic elements of tissue engineering. Microenvironment factors such as oxygen content and stress are key dynamic factors that can trigger the variations of angiogenesis. We may induce the formation of beneficial blood vessels and prevent the formation of pathological blood vessels by precisely regulating the microenvironment. In this chapter, we will elaborate the interaction between vascular endothelial cells and the extracellular microenvironment and summarize the influence of various microenvironment factors on angiogenesis. The finding that microenvironment factors play such a concerted role in angiogenesis suggests that incorporating microenvironment factors into tissue engineering might accelerate the development of novel therapeutics.

Keywords: Angiogenesis, Endothelial cell, Extracellular matrix, Microenvironment, Stiffness.

1. INTRODUCTION

Regenerative medicine technology is one of the most effective and promising methods to repair tissue trauma and function reconstruction. However, the researching results that can be transferred into clinical practice are extremely limited. The lack of nutrition and oxygen supply will lead to the failure of transplantation therapy. In this regard, researchers proposed several schemes that can promote the vascularization of tissue engineering products, including directly inoculating endothelial cells (ECs) onto scaffold materials to generate vascular structure before implantation. However, vascular structure in function was still very limited, for the reason that the scaffold material could not provide the appropriate microenvironment for the growth of related cells and new vessels.

---

[*] **Corresponding author Xiaoxiao Cai:** Sichuan University, West China School of Stomatology, China; E-mail: xcai@scu.edu.cn

Xiaoxiao Cai (Ed.)

Endothelial cells (ECs) are the vascular network's basic units, which constitute the lining of blood vessels throughout the body to the fine vascular networks which perfuse all tissues [1]. ECs line vessel walls as a monolayer by adhering to the basal lamina [2]. The vessels in different tissues and organs range in diameter from approximately 8 μm for capillaries to 2.5 cm for the aorta [3] The microvasculature system comprises only the intima and specialized SMCs, which offer both stability [4] and quiescence to capillaries [5]. As a result of their distinctive location between the volatile, intricate vascular wall and the flowing blood, ECs are subjected to forces from extracellular matrix (ECM) layers as well as surrounding tissues, such as muscle-mediated vessel contraction, physical forces from the residual hoop stress of the vessel wall, and physical inputs that originate from local ECM, which can impinge on the vessels [6]. Taken together, these microenvironment factors have a central role in managing physiological and pathological-related EC behavior, from the molecular to the tissue level.

EC is the basic unit of cell-based support therapy for cardiovascular and ischemic diseases. Guiding ECs biological behavior and optimizing angiogenesis by regulating the microenvironment factors has a great potential value in the treatment of vascular diseases, tumor diseases and so on. In this chapter, the effects of microenvironment factors on the angiogenesis in tissue engineering will be summarized.

2. EFFECT OF EXTRACELLULAR MATRIX ON ANGIOGENESIS IN TISSUE ENGINEERING

2.1. Properties of the Extracellular Matrix of Vessels

The ECM, composed of different functional proteins, is a highly dynamic structure which exists in all tissues and sustains continuous regulated remodeling. The remodeling is carried out by specific enzymes, such as metalloproteinases, accompanied by ECM degradation [7]. The ECM interacts with local cells to medicate various functions, such as proliferation, migration, and autophagy. ECM remodeling has a profound influence on the morphogenesis of cells and their functions [8]. Many pathological states, such as fibrosis or invasive cancer, are accompanied by dysregulation of the ECM structure, composition, abundance, and stiffness. It is imperative for researchers to obtain a thorough understanding of the way that ECM affects organ structure and function, and the way that ECM remodeling influences cell behavior and disease progression.

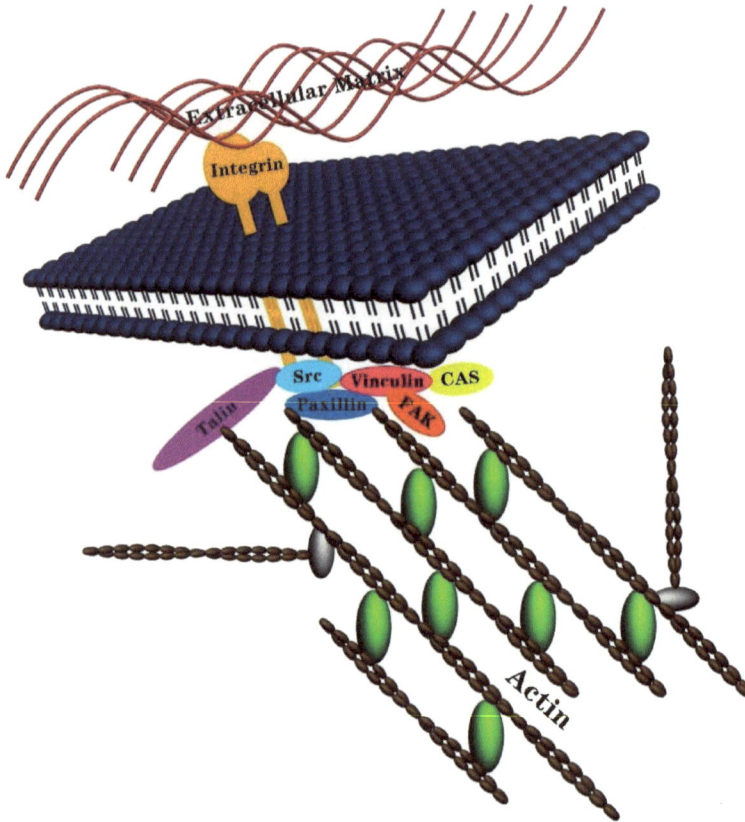

Fig. (1). FAs are ideal checkpoints which control inside-out and inwards transduction of biomechanical signals of the extracellular matrix.

The ECM of vessels is a network consisting of fibers and pores lined in an isotropic manner, ranging in size from the nanoscale (1–100 nm) to the submicron level (100–1000 nm) [9]. The local stiffness of the ECM has been reported to be between 5–140 kPa. The stiffness of the ECM plays a concerted role in vascular assembly and maintenance, even at the microvasculature level. In pathological conditions, ECM has been proved to have thickened in a number of tubes [10]. Moreover, ECs function is significantly affected by the elasticity of their local surrounding ECM, which provides matrix and stability for angiogenesis [11]. ECs have also been shown to secrete certain ECM components, which increase overall vessel rigidity in many pathological states [12]. Located basolaterally to all EC monolayers [13], the local ECM is a heterogeneous and complicated mixture of laminin; fibronectin; enactin; nidogen; collagen IV and V; the heparin-sulfate proteoglycan, perlacan; and so on [14 - 23]. A combination of biophysical and biochemical cues triggers the self-assembled layer of proteins which constitute the ECM, and the surrounding supportive cells (*e.g.* pericyte, PC), which are the principal constituents of the vascular microenvironment [24 - 26]. One of the components of the vessel ECM is the basal lamina, including the lamina lucida (15–65 nm thick) and the lamina densa (15–125 nm thick). In addition, some studies have reported the presence of a kind of lamina fibrorecticularis (2–15 μm) in some tissues [27, 28]. The ECM self assembles to offer support for the overlying ECs and delivers a variety of biochemical cues to them [15]. ECs anchor to, and interact with native ECM by focal adhesions (FAs), which comprise talin, integrin, α-actinin, vinculin, and many other proteins (Fig. **1**). While some other mechanisms might be involved, the formation and maintenance of FAs are mediated mainly by rhodopsin (RHO) and its related signaling cascade [29]. The ECM also acts as the reservoir of cytoactive growth factors, which include hormones and certain other molecules [30, 31].

2.2. Role of Matrix Stiffness in Regulating Endothelial Cells Behaviour

The local stiffness of the ECM is generally quantified using Young's modulus, which indicates the resistance to deformation of the basement membrane with which ECs interact [32 - 36]. The native stiffness of the ECM is distinct from the term vascular resistance, which refers to the variation in diameter of the tube under pressure [37 - 40]. As mentioned above, the stiffness (Young's modulus) of the local ECM of vasculature ranges from 5 kPa to 140 kPa [41 - 45]. Matrix stiffness also has an effect on the initiation and termination of vascular formation. During angiogenesis, ECM is gradually degraded by multiple matrix metalloproteinases (MMPs), restricting the existence of collagen IV. ECM, which contains a high content of laminin [46, 47], can also stretch out and induce cell migration [48, 49].

Matrix stiffness is critical to ECs' function and the network formation and maintenance of the vessels. The majority of *in vitro* stiffness works on EC behavior and angiogenesis have employed polydimethylsiloxane (PDMS), agarose, alginate, or polyacrylamide acid (PAA) [41]. Many of the related studies altered the stiffness by raising the concentration of the polymerized proteins [50, 51], adding cross-linkers [52], or modifying the cross-linking temperature or pH [53].

At the cytological level, matrix mechanics affect various EC functions. Zhang *et al.* showed that ECs tend to undergo an endothelial-to-mesenchymal transition and lose their typical endothelial phenotype when matrix stiffness increases. A decrease in stiffness favors the maintenance of the endothelial phenotype [54]. To a large extent, the cell spreading area, as one function of ECM stiffness tension, could be regarded as a measure of cell mechanosensitivity. On the whole, the more well spread a cell is, the more tension it receives from the matrix it adheres to [11]. Several studies indicated that ECs showed a more spread morphology on PA gels when the matrix stiffness increased [55]. When cultured on soft matrix, using PA gels derivatized with identical collagen concentrations, the ECs showed a more limited spreading area [56, 57]. EC cellular functions also change in response to matrix stiffness [58, 59]. Specifically, a study revealed that aortic MSCs also spread to a greater degree and have a more highly organized cytoskeleton on stiffer gels (≥ 66 kPa) compared with softer ones (≤ 22 kPa) [45]. Together with the increase in cell spreading area, EC migration becomes more vigorous with increasing matrix stiffness [60]. Another report suggested that proper vascular stiffness is crucial for the positive relationship between ECs and hemodynamic forces: While ECs are cultured on stiff-walled tubes and exposed to physiological pressure, the expression of endothelial nitric oxide synthase (eNOS) decreased compared with that in ECs on more compliant tubes [61].

Other than mechanical forces exerted on them, ECs also have a capability to impose forces on the ECM. ECs produce an equal amount of traction stress, which in turn remodels the ECM and assists migration [62]. Traction force is generated by actin filaments, bundles of myosins, stress fibers, and other proteins that anchor cells to the matrix through FAs [63]. Research on the physical characteristics of the EC cytoskeleton suggested that stress fibers function as viscoelastic cables, which are tensed through actomyosin interactions [64]. Meanwhile, the pre-extension of all stress fibers and cell endogenous forces hinge on the level of such actomyosin contractility [65]. Atomic force microscopy (AFM) has been applied to determine the moduli of ECs. Studies on human aortic ECs indicated point wise moduli of 1–2 kPa and up to 5 kPa in regions with and without actin stress fibers, respectively [66]. All the above evidence suggests that cell traction force generation, corresponding to matrix stiffness, serves as a regulator of many vital EC processes, in addition to mechano-sensing.

Evidences have indicated that traction force of ECs regulates endothelial cell–cell connectivity, as well as network formation. EC networks can be interrupted once the generation of traction forces is inhibited [67]. Another study showed that a reduction in physical resistance force of the matrix could accelerate EC network formation [68, 69]. ECs appear to become more morphologically similar once cell–cell connections are made [56]. This similarity of ECs might also induce the formation of the angiogenic network.

During vascular network formation, cell–cell connectivity is also closely associated with the optimization of the physical input [70]; on soft matrix, ECs seek out cell–cell connections and form networks to enhance the physical input [57]. By contrast, when presented with a rigid, physically-resistant matrix, in which vascular networks cannot form, the physical input might be optimized, creating a preference for EC–matrix interactions. On a sufficiently soft matrix, the traction forces exerted by ECs might literally generate a perceived stiffening on the matrix to draw cells together and organize them into vascular networks [60]. Meanwhile, during network formation, matrix stiffness might help to boost cell–cell connections. Studies have suggested that fibronectin polymerization is required for vascular network stabilization both in two-dimensional [57] and three-dimensional [71] structures.

2.3. Role of Matrix Stiffness in Regulating Angiogenesis

Various studies support the theory that mechanical stiffness of the microenvironment influences the formation of vascular networks [72]. ECs present on sufficiently soft matrix will self-assemble into vascular network structures in both two-dimensional and three-dimensional manners, independent of other exogenous biochemical cues [73]. Studies on two-dimensional network formation demonstrated that network formation by human umbilical vein endothelial cells (HUVECs) could be attenuated on rigid fibrin gels, and on matrigel that was made stiffer by polymerization with glycosylation or collagen I [74]. Bovine aortic ECs appeared to form an inactive network on stiffer collagen gels formed by increasing the collagen concentration [75]. In another two-dimensional system using PAA gels, the vessel network formation became more active on compliant gels. The PAA gel system can provide accurate control over matrix stiffness, while irrelevant of protein concentrations of the ECM applied to functionalize the gels [76]. One study revealed that ECs cultured on collagen-coated PAA hydrogels of diverse stiffness altered their morphology from a monolayer to a capillary-like phenotype when the stiffness of the gels decreased [68]. This phenomenon was further reproduced in studies carried out by Califano [57] and Saunders [77], showed that ECs could form networks only on soft PAA gels. A certain density of cells is required to assembly into vascular networks, and cells showed a limited spreading area and presented less active migration, indicating that on soft hydrogels, cell–cell connectivity is more dominant in comparing to cell–matrix adhesions, generating more stable vascular networks.

In three-dimensional settings, matrix stiffness is generally mediated by altering the collagen or fibrin density of the gels. It should be pointed out that altering the density could simultaneously change adhesion ligands and the gel structure. Even though in two-dimensional PAA settings, a relative explicit tendency toward fined tubular formation on soft gels was confirmed, this tendency in three-dimensional collagen/fibrin settings is ambiguous. However, the vast majority of works indicate that softer matrix induce vascular network formation. Vailhe *et al.*, for example, found that as the concentration of fibrin in three-dimensional hydrogels dropped from 2 to 0.5 mg/ml, the formation of a capillary-like network was enhanced. In relatively stiff gels, capillary-like networks cannot form at all [51, 78]. Kniazeva and colleagues used an *in vitro* model in which ECs were embedded in microcarrier beads then seeded into a three-dimensional fibrin structure. The results showed that longer vascular sprouts were found in gels with lower fibrin density [67]. In addition, Urech *et al.* regulated the physical properties of fibrin hydrogels by adding factor XIII to form more crosslinks, and found that a shorter, less capillary-like network formed when the cross-linker concentration and gel stiffness increased [79]. Edgar *et al.* used a computational model that indicated that higher collagen density created vascular networks with shorter lengths, decreased network interconnectivity, and fewer branching points [80]. Using a glycated collagen gel system, Francis-Sedlak *et al.* revealed that sprouting and angiogenesis of ECs were delayed in stiffer gels [81]. Furthermore, Shamloo *et al.* showed that an intermediate density of collagen (1.2–1.9 mg/ml) generated the best vessel sprouting effect [82]. Bovine pulmonary aortic ECs delivered with microcarrier beads distributed into fibrin hyrdogels showed less capillary-like formation as the stiffness increased [83], and the same trend was shown in a HUVEC model [73].

In summary, the above data suggested that more compliant matrix tend to facilitate angiogenesis. However, the regulatory effect of matrix stiffness on the vascular network formation in three-dimensional gel models remains undefined. Improved three-dimensional systems to regulate matrix stiffness and more systematic studies are required to obtain more definite conclusions.

2.4. Underlying Mechanism of Matrix Stiffness on Angiogenesis

2.4.1. Mechanosensing and Mechanotransduction

Vessel sprouting refers to the proliferation, migration, and invasion of ECs, which extend sprouts to construct vascular structures. The events above include cytoskeleton-related proteins, as well as actomyosin contractility, and are under the regulation of the upstream effectors Rho and Rho-associated protein kinase (Rock) signaling pathways [84]. The small GTPase family, which regulates actomyosin tension, regulates many signals during vascular network formation by assisting ECs to sense and respond to matrix stiffness [85].

Separate lines of evidence suggest a key role for matrix stiffness in the mediation of the Rho-GTPase signaling balance, with consequences for the intercellular force balance and the integrity of endothelial intercellular junctions [86 - 88]. More attention should be paid on the dual behavior of Rho-GTPases, which could affect the endothelial barrier status. Kniazeva *et al.* investigated the connection between actin-mediated contractility and the small GTPase family by applying a series of force inhibitors to regulate the growth of new vessels. They observed that these inhibitors inhibited new vessel sprout and vessel maturation, causing in thinning and dissociation of the lumens [67]. Moreover, controlling the microtubule-depolymerizing agents brought about quick collapse of vascular networks, indicating that microtubules are key cytoskeleton proteins that play a vital role in the maintenance of vessel integrity [89]. De-polymerization was shown to activate Rho signaling, leading to vascular network collapse, mostly because of the enhanced tension on actin stress fibers by Rho-dependent or Rock-dependent stimulation of actomyosin motors [89]. Increasing numbers of studies have focused on the influence of ECM stiffness on the cytoskeleton and FAs. Evidence showed that soft collagen gels downregulated integrin(β1) activation as well as focal adhesion kinase Y397 phosphorylation [90]. Moreover, cells exposed to higher extracellular pressure increased the focal adhesion kinase phosphorylation level at Y576 and Y397, in an α-actinin-1 phosphorylation-dependent manner, suggesting that FAK plays a crucial role in the mechanotransduction initiated by matrix stiffness [91]. As for ECs, when grown in softer gels, they exhibit less organized stress fibers compared with cells cultured in stiffer hydrogels [68, 92]. Furthermore, the protein levels of integrin (α2), talin, vinculin, actin, and α-actinin were altered by varying ECM stiffness [68]. Under circumstances in which ECs can assemble into vascular network structures (on soft gels), notable reductions in the levels of adhesion-related and cytoskeleton-related proteins were observed compared with those in ECs forming a monolayer on stiffer gels [93].

Myosin II is generally regarded as a crucial effector of EC branching. Fischer showed myosin II is mostly located at the cortex of EC tip cells while functions to restrain cell branching. By contrast, native depletion of myosin II can trigger branching [94]. At locations of intense Myosin II staining, the majority of EC protrusions were retracted because of the strong tension. On the other hand, in ECM softening regions, contractility is downregulated and Myosin II is decreased further, which may assist the protrusion progression. In compliance ECM microenvironments, ECs sprout many branches, and they cannot change direction. In stiffer microenvironments, by contrast, fewer branches are formed and they keep extending in the pre-existing branch direction. A connection between matrix stiffness and ECs branching has been proposed, and the proposed models provide a comprehensive understanding of the relationship between matrix stiffness, cytoskeleton, and Rho-GTPase signals to mediate endothelial sprouting behavior and vascular network formation.

2.4.2. Expression of Functional Proteins and Growth Factors

Extracellular stiffness plays a key role in mediating EC behavior and angiogenesis; however, we require a deeper understanding of the precise method by which this mechanical factor acts (Fig. **2**). ECs behavior can be affected by matrix stiffness, indicating that VEGF activity may be regulated by mechanical cellular microenvironment [95 - 97].

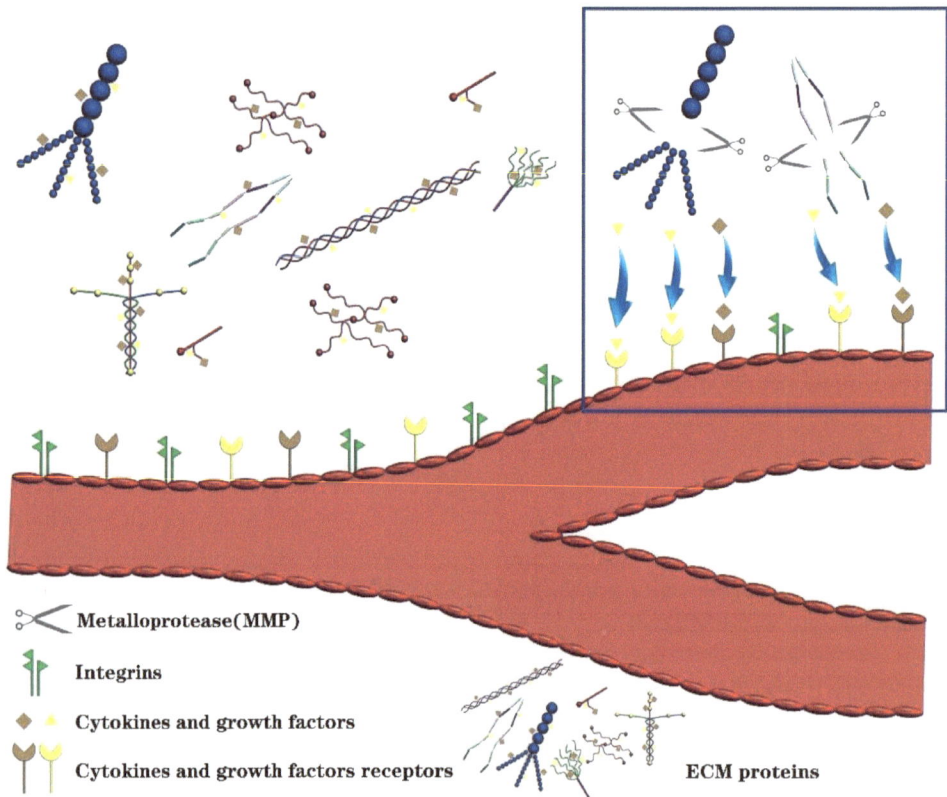

Fig. (2). Schematic representation of key ECM components involved in angiogenesis.

Shamloo *et al.* revealed that mechanical properties, such as substrate density, not only affect ECs directly, but can also influence the secretion of angiogenesis-related growth factors, such as VEGF [82]. In a collagen-coated PAA hydrogel model, HUVECs were exposed to a microenvironment that mimicked the stiffness properties of collagenous bone; gene and protein levels, and cell proliferation, were estimated. The gene and protein expression levels and the proliferation of HUVECs were not significantly influenced on 3 kPa compared with 30 kPa PAA gels. Although most of the gene transcript levels and proliferation remained constant, a decreased expression level of the functional VEGF receptor-2 was noted in the stiffer gels [98]. Evidence from *in vivo* studies suggests that a decrease in matrix stiffness correlates with neo-vessel formation and MMPs levels [99].

In particular, in the vascular endothelium, soft matrix tend to induce NO synthesis and are thus more likely to facilitate angiogenesis and increase tissue perfusion [100 - 102]. To date, two mechanisms have been reported to explain the way that soft matrix boost eNOS activity. First, eNOS activity can be activated by G-actin, while it is attenuated by F-actin [103]. Softer matrix stiffness leads to F-actin depolymerization and therefore stimulates NO release [104]. Second, a compliant matrix might render ECs more susceptible to blood flow shear stress. In addition, angiogenesis and vascular maintenance require the elaboration of endothelium-derived nitric oxide [105, 106].

Collectively, the above data indicate that mechanical factors, such as matrix stiffness, are key regulators of EC behavior and vascular network formation. Revealing the connection between matrix stiffness and angiogenesis is important to understand important processes, such as wound healing, organ transplantation, and the prevention of multiple disorders, especially tumor growth.

3. EFFECT OF FLUID SHEAR FORCE ON ANGIOGENESIS

Fluid shear force is a widely existing force in the blood system. It is a tangential force parallel to the surface of blood vessel formed by the friction between blood and the vessel wall. The ECs constitutes the inner layer of blood vessel and can continuously sense the fluid shear force. First, evidences showed that fluid shear force can regulate the migration of ECs, especially in the large vessels with strong shear force. The regulatory effect on cell migration was even stronger than chemotaxis [107]. In addition, fluid shear force can also increase the activity of a variety of enzymes and regulate the phosphorylation of ECs signalling proteins [108, 109]. Some *in vitro* experiments have confirmed that ECs that exposed to high shear stress showed a state of accelerated proliferation and ECM remodelling [110 - 114]. It has also been suggested that the process of vascular remodelling is mainly mediated by fluid shear stress [115]. In the process of wound healing, fluid shear stress can promote the production of microcirculation network by inducing the formation of lamellar pseudopodia in ECs [116]. In addition, the shear stress can also affect the cell axis, which was shown that the long axis of cell arrangement will extend along the direction of blood flow. The cytoskeleton was also polarized under the action of fluid shear force [117].

There is also evidence that the arteriovenous differentiation is mediated by fluid shear stress [118]. A number of experiments and clinical data have proved that fluid shear stress induces arterial differentiation of endothelial progenitor cells [119]. A study of hypertensive rat model showed that compared with normal blood pressure group, the characterization of blood vessels and arteries in hypertensive group was significantly enhanced [120]. The arterial differentiation trend of the implanted blood vessel grafts was also a typical example. In that experiment model, the vein tissue was implanted and connected to the artery. The fluid shear force arterialized the implanted vein, which proved that the biomechanical factor has a strong effect on the angiomorphism of the artery [121].

4. EFFECT OF MICROPATTERNS OF SUBSTRATE MATERIALS ON ANGIOGENESIS

In tissue engineering, many researches are focused on the microstructure of vascularized materials. The new material manufacturing technology was used to design morphology of microenvironment for the vascular related cells, so as to regulate the biological cell behaviour and the formation of blood vessels.

More optimized micropatterns of substrate materials were explored by researchers to facilitate the functional vascularization. One strategy is to prepare scaffolds or hydrogels with hollow channels, which can inoculate vascular ECs and form designed vessels in the channel. Various methods have been reported to prepare such scaffold materials with hollow channels, such as three-dimensional fiber deposition technology [122, 123], electrospinning scaffold materials [124], laser drilling [125], and using silicon mold to copy micro-patterns [126, 127]. In recent years, the preparation of tubular channels in hydrogels can provide a better microenvironment for cells. ECs can cluster and sprout branches in the hydrogel structure [128, 129], and then construct a three-dimensional vascular network [130 - 132]. A study has shown that after the implantation into a 20 mm wide channel, the ECs formed a millimetre-sized capillary and further supported a functional tissue [133].

In addition to building the channel structures in scaffold materials, the formation of blood vessels can also be more precisely controlled by designing appropriate micropattern geometry. In three-dimensional model, patterned materials can be obtained by cell slicing technology [134], light patterning technology [135] or biological printing technology [136, 137]. Complex vascular networks can be created in multiple dimensions by micropatterning ECs. At present, much effort has been made to improve the design of vascular scaffold by simulating ECM micropattern [138 - 143]. Studies have concluded that submicron (micrograph) to nanometer (micrograph) morphological characteristics may affect adhesion, migration and proliferation of ECs [144 - 150], and regulate arterial-venous differentiation [151 - 153]. Chaturvedi *et al.* [154]. showed that after being implanted into mice, patterned materials built up neovascularization system that connected with the autologous vascular system to achieve effective perfusion. In clinical application, micropatterns factor has been taken into account in the design of nitinol and titanium oxide alloy stent [145].

5. EFFECT OF CELL LOADING STRESS ON ANGIOGENESIS

The theory that mechanical loading can regulate the self-renewal and differentiation of stem cells has become a widespread consensus [155]. There are many phenomena that can produce loading stress, such as muscle contraction, heart beating and so on. For example, vascular ECs can remodel the vascular wall in the case of persistent hypertension [156]. Moreover, under the stretching and compression cycling force for 60 time per minute, the tyrosine phosphorylation of bovine aortic endothelial cells was triggered and focal adhesions were remodelled [157].

6. EFFECT OF NITRIC OXIDE LEVEL ON ANGIOGENESIS

Nitric oxide (NO) can regulate the proliferation of ECs, trigger vasoactive substances, and play a role in maintaining the integrity of vascular endothelium under physiological and pathological conditions [158 - 160]. The most direct influencing factor of endothelial dysfunction is the decrease of NO level, which will subsequently cause a significant decrease of eNOS level [161, 162].

CONCLUDING REMARKS

The significance of angiogenesis for the clinical applicability of regenerative medicine has resulted in many approaches being explored to include an organized functional vascular network for tissue constructs. Although some molecular regulators of matrix stiffness still need to be identified, increasing evidence suggests that hemodynamic factors such as matrix stiffness are crucial determinants of the appearance and the structure of the neovascularization. Matrix stiffness may be considered as a complementary tool for a better control of angiogenesis, or as a strategy to stabilize an existing organized vascular network.

CONSENT FOR PUBLICATION

Not applicable.

CONFLICT OF INTEREST

The authors confirm that the contents of this chapter have no conflict of interest.

ACKNOWLEDGEMENTS

This study was supported by National Key R&D Program of China (2019YFA0110600) and National Natural Science Foundation of China (81970986, 81771125).

REFERENCES

[1]   Kalluri R. Basement membranes: structure, assembly and role in tumour angiogenesis. Nat Rev Cancer 2003; 3(6): 422-33.
        [http://dx.doi.org/10.1038/nrc1094] [PMID: 12778132]

[2]   Gössl M, Rosol M, Malyar NM, *et al.* Functional anatomy and hemodynamic characteristics of vasa vasorum in the walls of porcine coronary arteries. Anat Rec A Discov Mol Cell Evol Biol 2003; 272(2): 526-37.
        [http://dx.doi.org/10.1002/ar.a.10060] [PMID: 12740947]

[3]   von Tell D, Armulik A, Betsholtz C. Pericytes and vascular stability. Exp Cell Res 2006; 312(5): 623-9.
        [http://dx.doi.org/10.1016/j.yexcr.2005.10.019] [PMID: 16303125]

[4]   Antonelli-Orlidge A, Saunders KB, Smith SR, D'Amore PA. An activated form of transforming growth factor β is produced by cocultures of endothelial cells and pericytes. Proc Natl Acad Sci USA 1989; 86(12): 4544-8.
        [http://dx.doi.org/10.1073/pnas.86.12.4544] [PMID: 2734305]

[5]   Burton AC. Relation of structure to function of the tissues of the wall of blood vessels. Physiol Rev 1954; 34(4): 619-42.
        [http://dx.doi.org/10.1152/physrev.1954.34.4.619] [PMID: 13215088]

[6]   Sheriff D. Point: The muscle pump raises muscle blood flow during locomotion. J Appl Physiol 2005; 99(1): 371-2.
        [http://dx.doi.org/10.1152/japplphysiol.00381.2005] [PMID: 16036908]

[7]   Watt FM, Huck WT. Role of the extracellular matrix in regulating stem cell fate. Nat Rev Mol Cell Biol 2013; 14(8): 467-73.
        [http://dx.doi.org/10.1038/nrm3620] [PMID: 23839578]

[8]   Du J, Chen X, Liang X, *et al.* Integrin activation and internalization on soft ECM as a mechanism of induction of stem cell differentiation by ECM elasticity. Proc Natl Acad Sci USA 2011; 108(23): 9466-71.
        [http://dx.doi.org/10.1073/pnas.1106467108] [PMID: 21593411]

[9]   Wood JA, Liliensiek SJ, Russell P, Nealey PF, Murphy CJ. Biophysical cueing and vascular endothelial cell behavior. Materials (Basel) 2010; 3: 1620-39.
        [http://dx.doi.org/10.3390/ma3031620]

[10]   Simionescu M, Antohe F. Functional ultrastructure of the vascular endothelium: changes in various pathologies. Handb Exp Pharmacol 2006; 176(176 Pt 1): 41-69.
        [http://dx.doi.org/10.1007/3-540-32967-6_2] [PMID: 16999216]

[11]   Chicurel ME, Chen CS, Ingber DE. Cellular control lies in the balance of forces. Curr Opin Cell Biol 1998; 10(2): 232-9.
        [http://dx.doi.org/10.1016/S0955-0674(98)80145-2] [PMID: 9561847]

[12] Last JA, Liliensiek SJ, Nealey PF, Murphy CJ. Determining the mechanical properties of human corneal basement membranes with atomic force microscopy. J Struct Biol 2009; 167(1): 19-24.
[http://dx.doi.org/10.1016/j.jsb.2009.03.012] [PMID: 19341800]

[13] LeBleu VS, Macdonald B, Kalluri R. Structure and function of basement membranes. Exp Biol Med (Maywood) 2007; 232(9): 1121-9.
[http://dx.doi.org/10.3181/0703-MR-72] [PMID: 17895520]

[14] Candiello J, Balasubramani M, Schreiber EM, *et al.* Biomechanical properties of native basement membranes. FEBS J 2007; 274(11): 2897-908.
[http://dx.doi.org/10.1111/j.1742-4658.2007.05823.x] [PMID: 17488283]

[15] Liliensiek SJ, Nealey P, Murphy CJ. Characterization of endothelial basement membrane nanotopography in rhesus macaque as a guide for vessel tissue engineering. Tissue Eng Part A 2009; 15(9): 2643-51.
[http://dx.doi.org/10.1089/ten.tea.2008.0284] [PMID: 19207042]

[16] Discher DE, Janmey P, Wang YL. Tissue cells feel and respond to the stiffness of their substrate. Science 2005; 310(5751): 1139-43.
[http://dx.doi.org/10.1126/science.1116995] [PMID: 16293750]

[17] Abrams G, Teixeira A, Nealey P, Murphy C. Effects of Substratum Topography on Cell Behavior. Biomimetic Materials and Design 2002; 90-137.
[http://dx.doi.org/10.1201/9780203908976.ch4]

[18] Brody S, Anilkumar T, Liliensiek S, Last JA, Murphy CJ, Pandit A. Characterizing nanoscale topography of the aortic heart valve basement membrane for tissue engineering heart valve scaffold design. Tissue Eng 2006; 12(2): 413-21.
[http://dx.doi.org/10.1089/ten.2006.12.413] [PMID: 16548699]

[19] Timpl R. Macromolecular organization of basement membranes. Curr Opin Cell Biol 1996; 8(5): 618-24.
[http://dx.doi.org/10.1016/S0955-0674(96)80102-5] [PMID: 8939648]

[20] Kolega J, Manabe M, Sun TT. Basement membrane heterogeneity and variation in corneal epithelial differentiation. Differentiation 1989; 42(1): 54-63.
[http://dx.doi.org/10.1111/j.1432-0436.1989.tb00607.x] [PMID: 2695378]

[21] Ekblom P, Timpl R. Cell-to-cell contact and extracellular matrix. A multifaceted approach emerging. Curr Opin Cell Biol 1996; 8(5): 599-601.
[http://dx.doi.org/10.1016/S0955-0674(96)80099-8] [PMID: 8939665]

[22] Merker HJ. Morphology of the basement membrane. Microsc Res Tech 1994; 28(2): 95-124.
[http://dx.doi.org/10.1002/jemt.1070280203] [PMID: 8054667]

[23] Yurchenco P D, O'Rear J. Supramolecular organization of basement membranes. Molecular and Cellular Aspects of Basement Membrane 1993: 19–47.
[http://dx.doi.org/10.1016/B978-0-12-593165-6.50008-5]

[24] Le Saux O, Teeters K, Miyasato S, *et al.* The role of caveolin-1 in pulmonary matrix remodeling and mechanical properties. Am J Physiol Lung Cell Mol Physiol 2008; 295(6): L1007-17.
[http://dx.doi.org/10.1152/ajplung.90207.2008] [PMID: 18849439]

[25] Godier AF, Marolt D, Gerecht S, Tajnsek U, Martens TP, Vunjak-Novakovic G. Engineered microenvironments for human stem cells. Birth Defects Res C Embryo Today 2008; 84(4): 335-47.
[http://dx.doi.org/10.1002/bdrc.20138] [PMID: 19067427]

[26] Figallo E, Cannizzaro C, Gerecht S, *et al.* Micro-bioreactor array for controlling cellular microenvironments. Lab Chip 2007; 7(6): 710-9.
[http://dx.doi.org/10.1039/b700063d] [PMID: 17538712]

[27] Tanner GA, Evan AP, Williams JC. Reply to Miner, Ajp. Ren Physiol 2009; 297: F551.

[http://dx.doi.org/10.1152/ajprenal.00259.2009]

[28]   Møbjerg N, Jespersen A, Wilkinson M. Morphology of the kidney in the West African caecilian, Geotrypetes seraphini (Amphibia, Gymnophiona, Caeciliidae). J Morphol 2004; 262(2): 583-607. [http://dx.doi.org/10.1002/jmor.10244] [PMID: 15376276]

[29]   Ridley AJ, Hall A. The small GTP-binding protein rho regulates the assembly of focal adhesions and actin stress fibers in response to growth factors. Cell 1992; 70(3): 389-99. [http://dx.doi.org/10.1016/0092-8674(92)90163-7] [PMID: 1643657]

[30]   Bentley E, Murphy CJ. Topical therapeutic agents that modulate corneal wound healing. Vet Clin North Am Small Anim Pract 2004; 34(3): 623-38. [http://dx.doi.org/10.1016/j.cvsm.2003.12.006] [PMID: 15110975]

[31]   Davis GE, Senger DR. Endothelial extracellular matrix: biosynthesis, remodeling, and functions during vascular morphogenesis and neovessel stabilization. Circ Res 2005; 97(11): 1093-107. [http://dx.doi.org/10.1161/01.RES.0000191547.64391.e3] [PMID: 16306453]

[32]   Ebenstein DM, Pruitt LA. Nanoindentation of soft hydrated materials for application to vascular tissues. J Biomed Mater Res A 2004; 69(2): 222-32. [http://dx.doi.org/10.1002/jbm.a.20096] [PMID: 15057995]

[33]   Oie T, Murayama Y, Fukuda T, *et al.* Local elasticity imaging of vascular tissues using a tactile mapping system. J Artif Organs 2009; 12(1): 40-6. [http://dx.doi.org/10.1007/s10047-008-0440-5] [PMID: 19330504]

[34]   Lundkvist A, Lilleodden E, Siekhaus W, Kinney J, Balooch LPM. Viscoelastic Properties of Healthy Human Artery Measured in Saline Solution by AFM-Based Indentation Technique. MRS Online Proceeding Library Archive 1996; p. 436.

[35]   Ebenstein DM, Pruitt LA. Nanoindentation of biological materials. Nano Today 2006; 1: 26-33. [http://dx.doi.org/10.1016/S1748-0132(06)70077-9]

[36]   Cao Y, Ma D, Raabe D. The use of flat punch indentation to determine the viscoelastic properties in the time and frequency domains of a soft layer bonded to a rigid substrate. Acta Biomater 2009; 5(1): 240-8. [http://dx.doi.org/10.1016/j.actbio.2008.07.020] [PMID: 18722168]

[37]   Yang J, Motlagh D, Webb AR, Ameer GA. Novel biphasic elastomeric scaffold for small-diameter blood vessel tissue engineering. Tissue Eng 2005; 11(11-12): 1876-86. [http://dx.doi.org/10.1089/ten.2005.11.1876] [PMID: 16411834]

[38]   Zhang G, Varkey M, Wang Z, Xie B, Hou R, Atala A. ECM concentration and cell-mediated traction forces play a role in vascular network assembly in 3D bioprinted tissue. Biotechnol Bioeng 2020; 117(4): 1148-58. [http://dx.doi.org/10.1002/bit.27250] [PMID: 31840798]

[39]   Engler AJ, Griffin MA, Sen S, Bönnemann CG, Sweeney HL, Discher DE. Myotubes differentiate optimally on substrates with tissue-like stiffness: pathological implications for soft or stiff microenvironments. J Cell Biol 2004; 166(6): 877-87. [http://dx.doi.org/10.1083/jcb.200405004] [PMID: 15364962]

[40]   Hozhabr M, Zhou C, *et al.* Mechanical contribution of vascular smooth muscle cells in the tunica media of artery. Nanotechnol Rev 2019; 8(1): 50-60. [http://dx.doi.org/10.1515/ntrev-2019-0005]

[41]   Brown XQ, Ookawa K, Wong JY. Evaluation of polydimethylsiloxane scaffolds with physiologically-relevant elastic moduli: interplay of substrate mechanics and surface chemistry effects on vascular smooth muscle cell response. Biomaterials 2005; 26(16): 3123-9. [http://dx.doi.org/10.1016/j.biomaterials.2004.08.009] [PMID: 15603807]

[42]   Lee JC, Discher DE. Deformation-enhanced fluctuations in the red cell skeleton with theoretical relations to elasticity, connectivity, and spectrin unfolding. Biophys J 2001; 81(6): 3178-92.

[http://dx.doi.org/10.1016/S0006-3495(01)75954-1] [PMID: 11720984]

[43] Engler A, Bacakova L, Newman C, Sheehan M. Mechanical role of cytoskeletal components in vascular smooth muscle cell adhesion *in vitro*. Proceedings of the IEEE 28th Annual Northeast Bioengineering Conference (IEEE Cat No02CH37342). Philadelphia, PA, USA 2002, pp. 23-24.
[http://dx.doi.org/10.1109/NEBC.2002.999446]

[44] Engler AJ, Sen S, Sweeney HL, Discher DE. Matrix elasticity directs stem cell lineage specification. Cell 2006; 126(4): 677-89.
[http://dx.doi.org/10.1016/j.cell.2006.06.044] [PMID: 16923388]

[45] Engler A, Bacakova L, Newman C, Hategan A, Griffin M, Discher D. Substrate compliance *versus* ligand density in cell on gel responses. Biophys J 2004; 86(1 Pt 1): 617-28.
[http://dx.doi.org/10.1016/S0006-3495(04)74140-5] [PMID: 14695306]

[46] Jacot JG, Dianis S, Schnall J, Wong JY. A simple microindentation technique for mapping the microscale compliance of soft hydrated materials and tissues. J Biomed Mater Res A 2006; 79(3): 485-94.
[http://dx.doi.org/10.1002/jbm.a.30812] [PMID: 16779854]

[47] Engler AJ, Richert L, Wong JY, Picart C, Discher DE. Surface probe measurements of the elasticity of sectioned tissue, thin gels and polyelectrolyte multilayer films: Correlations between substrate stiffness and cell adhesion. Surf Sci 2004; 570: 142-54.
[http://dx.doi.org/10.1016/j.susc.2004.06.179]

[48] Form DM, Pratt BM, Madri JA. Endothelial cell proliferation during angiogenesis. *In Vitro* modulation by basement membrane components. Lab Invest 1986; 55(5): 521-30.
[PMID: 2430138]

[49] Ingber DE. Mechanical signaling and the cellular response to extracellular matrix in angiogenesis and cardiovascular physiology. Circ Res 2002; 91(10): 877-87.
[http://dx.doi.org/10.1161/01.RES.0000039537.73816.E5] [PMID: 12433832]

[50] Vernon RB, Lara SL, Drake CJ, *et al.* Organized type I collagen influences endothelial patterns during "spontaneous angiogenesis *in vitro*": planar cultures as models of vascular development. *In Vitro* Cell Dev Biol Anim 1995; 31(2): 120-31.
[http://dx.doi.org/10.1007/BF02633972] [PMID: 7537585]

[51] Vailhé B, Ronot X, Tracqui P, Usson Y, Tranqui L. *In vitro* angiogenesis is modulated by the mechanical properties of fibrin gels and is related to α(v)β3 integrin localization. *In Vitro* Cell Dev Biol Anim 1997; 33(10): 763-73.
[http://dx.doi.org/10.1007/s11626-997-0155-6] [PMID: 9466681]

[52] Georges PC, Janmey PA. Cell type-specific response to growth on soft materials. J Appl Physiol 2005; 98(4): 1547-53.
[http://dx.doi.org/10.1152/japplphysiol.01121.2004] [PMID: 15772065]

[53] Yamamura N, Sudo R, Ikeda M, Tanishita K. Effects of the mechanical properties of collagen gel on the in vitro formation of microvessel networks by endothelial cells. Tissue Eng 2007; 13(7): 1443-53.
[http://dx.doi.org/10.1089/ten.2006.0333] [PMID: 17518745]

[54] Zhang H, Chang H, Wang LM, *et al.* Effect of polyelectrolyte film stiffness on endothelial cells during endothelial-to-mesenchymal transition. Biomacromolecules 2015; 16(11): 3584-93.
[http://dx.doi.org/10.1021/acs.biomac.5b01057] [PMID: 26477358]

[55] Reinhart-King CA. Endothelial cell adhesion and migration. Methods Enzymol 2008; 443: 45-64.
[http://dx.doi.org/10.1016/S0076-6879(08)02003-X] [PMID: 18772010]

[56] Yeung T, Georges PC, Flanagan LA, *et al.* Effects of substrate stiffness on cell morphology, cytoskeletal structure, and adhesion. Cell Motil Cytoskeleton 2005; 60(1): 24-34.
[http://dx.doi.org/10.1002/cm.20041] [PMID: 15573414]

[57] Califano JP, Reinhart-King CA. A balance of substrate mechanics and matrix chemistry regulates

endothelial cell network assembly. Cell Mol Bioeng 2008; 1: 122.
[http://dx.doi.org/10.1007/s12195-008-0022-x]

[58]   Ghajar CM, Chen X, Harris JW, *et al.* The effect of matrix density on the regulation of 3-D capillary morphogenesis. Biophys J 2008; 94(5): 1930-41.
[http://dx.doi.org/10.1529/biophysj.107.120774] [PMID: 17993494]

[59]   Griffith LG, Swartz MA. Capturing complex 3D tissue physiology *in vitro*. Nat Rev Mol Cell Biol 2006; 7(3): 211-24.
[http://dx.doi.org/10.1038/nrm1858] [PMID: 16496023]

[60]   Reinhart-King CA, Dembo M, Hammer DA. Cell-cell mechanical communication through compliant substrates. Biophys J 2008; 95(12): 6044-51.
[http://dx.doi.org/10.1529/biophysj.107.127662] [PMID: 18775964]

[61]   Peng X, Haldar S, Deshpande S, Irani K, Kass DA. Wall stiffness suppresses Akt/eNOS and cytoprotection in pulse-perfused endothelium. Hypertension 2003; 41(2): 378-81.
[http://dx.doi.org/10.1161/01.HYP.0000049624.99844.3D] [PMID: 12574111]

[62]   Cynthia A, King R, Dembo R, *et al.* Endothelial cell traction forces on RGD-derivatized polyacrylamide substrata†. Langmuir 2003; 19: 1573-9.
[http://dx.doi.org/10.1021/la026142j]

[63]   Lu L, Oswald SJ, Ngu H, Yin FC. Mechanical properties of actin stress fibers in living cells. Biophys J 2008; 95(12): 6060-71.
[http://dx.doi.org/10.1529/biophysj.108.133462] [PMID: 18820238]

[64]   Kumar S, Maxwell IZ, Heisterkamp A, *et al.* Viscoelastic retraction of single living stress fibers and its impact on cell shape, cytoskeletal organization, and extracellular matrix mechanics. Biophys J 2006; 90(10): 3762-73.
[http://dx.doi.org/10.1529/biophysj.105.071506] [PMID: 16500961]

[65]   Lu L, Feng Y, Hucker WJ, Oswald SJ, Longmore GD, Yin FCP. Actin stress fiber pre-extension in human aortic endothelial cells. Cell Motil Cytoskeleton 2008; 65(4): 281-94.
[http://dx.doi.org/10.1002/cm.20260] [PMID: 18200567]

[66]   Costa KD, Sim AJ, Yin FC. Non-Hertzian approach to analyzing mechanical properties of endothelial cells probed by atomic force microscopy. J Biomech Eng 2006; 128(2): 176-84.
[http://dx.doi.org/10.1115/1.2165690] [PMID: 16524328]

[67]   Kniazeva E, Putnam AJ. Endothelial cell traction and ECM density influence both capillary morphogenesis and maintenance in 3-D. Am J Physiol Cell Physiol 2009; 297(1): C179-87.
[http://dx.doi.org/10.1152/ajpcell.00018.2009] [PMID: 19439531]

[68]   Deroanne CF, Lapiere CM, Nusgens BV. In vitro tubulogenesis of endothelial cells by relaxation of the coupling extracellular matrix-cytoskeleton. Cardiovasc Res 2001; 49(3): 647-58.
[http://dx.doi.org/10.1016/S0008-6363(00)00233-9] [PMID: 11166278]

[69]   Ingber DE, Folkman J. Mechanochemical switching between growth and differentiation during fibroblast growth factor-stimulated angiogenesis in vitro: role of extracellular matrix. J Cell Biol 1989; 109(1): 317-30.
[http://dx.doi.org/10.1083/jcb.109.1.317] [PMID: 2473081]

[70]   Guo WH, Frey MT, Burnham NA, Wang YL. Substrate rigidity regulates the formation and maintenance of tissues. Biophys J 2006; 90(6): 2213-20.
[http://dx.doi.org/10.1529/biophysj.105.070144] [PMID: 16387786]

[71]   Zhou X, Rowe RG, Hiraoka N, *et al.* Fibronectin fibrillogenesis regulates three-dimensional neovessel formation. Genes Dev 2008; 22(9): 1231-43.
[http://dx.doi.org/10.1101/gad.1643308] [PMID: 18451110]

[72]   Lavalley D J, Reinhart-King C A. Matrix stiffening in the formation of blood vessels Advance in regenerative biology 2014 2014.

[http://dx.doi.org/10.3402/arb.v1.25247]

[73] Sieminski AL, Hebbel RP, Gooch KJ. The relative magnitudes of endothelial force generation and matrix stiffness modulate capillary morphogenesis in vitro. Exp Cell Res 2004; 297(2): 574-84.
[http://dx.doi.org/10.1016/j.yexcr.2004.03.035] [PMID: 15212957]

[74] Kuzuya M, Satake S, Miura H, Hayashi T, Iguchi A. Inhibition of endothelial cell differentiation on a glycosylated reconstituted basement membrane complex. Exp Cell Res 1996; 226(2): 336-45.
[http://dx.doi.org/10.1006/excr.1996.0234] [PMID: 8806437]

[75] Kuzuya M, Satake S, Ai S, *et al.* Inhibition of angiogenesis on glycated collagen lattices. Diabetologia 1998; 41(5): 491-9.
[http://dx.doi.org/10.1007/s001250050937] [PMID: 9628264]

[76] Tse JR, Engler AJ. Preparation of hydrogel substrates with tunable mechanical properties. Curr Protoc Cell Biol 2010; Chapter 10.

[77] Saunders RL, Hammer DA. Assembly of human umbilical vein endothelial cells on compliant hydrogels. Cell Mol Bioeng 2010; 3(1): 60-7.
[http://dx.doi.org/10.1007/s12195-010-0112-4] [PMID: 21754971]

[78] Vailhé B, Lecomte M, Wiernsperger N, Tranqui L. The formation of tubular structures by endothelial cells is under the control of fibrinolysis and mechanical factors. Angiogenesis 1998; 2(4): 331-44.
[http://dx.doi.org/10.1023/A:1009238717101] [PMID: 14517453]

[79] Urech L, Bittermann AG, Hubbell JA, Hall H. Mechanical properties, proteolytic degradability and biological modifications affect angiogenic process extension into native and modified fibrin matrices *in vitro*. Biomaterials 2005; 26(12): 1369-79.
[http://dx.doi.org/10.1016/j.biomaterials.2004.04.045] [PMID: 15482824]

[80] Edgar LT, Underwood CJ, Guilkey JE, Hoying JB, Weiss JA. Extracellular matrix density regulates the rate of neovessel growth and branching in sprouting angiogenesis. PLoS One 2014; 9(1): e85178.
[http://dx.doi.org/10.1371/journal.pone.0085178] [PMID: 24465500]

[81] Francis-Sedlak ME, Moya ML, Huang JJ, *et al.* Collagen glycation alters neovascularization *in vitro* and *in vivo*. Microvasc Res 2010; 80(1): 3-9.
[http://dx.doi.org/10.1016/j.mvr.2009.12.005] [PMID: 20053366]

[82] Shamloo A, Heilshorn SC. Matrix density mediates polarization and lumen formation of endothelial sprouts in VEGF gradients. Lab Chip 2010; 10(22): 3061-8.
[http://dx.doi.org/10.1039/c005069e] [PMID: 20820484]

[83] Nehls V, Herrmann R. The configuration of fibrin clots determines capillary morphogenesis and endothelial cell migration. Microvasc Res 1996; 51(3): 347-64.
[http://dx.doi.org/10.1006/mvre.1996.0032] [PMID: 8992233]

[84] Mammoto A, Mammoto T, Ingber DE. Rho signaling and mechanical control of vascular development. Curr Opin Hematol 2008; 15(3): 228-34.
[http://dx.doi.org/10.1097/MOH.0b013e3282fa7445] [PMID: 18391790]

[85] Ghosh K, Thodeti CK, Dudley AC, Mammoto A, Klagsbrun M, Ingber DE. Tumor-derived endothelial cells exhibit aberrant Rho-mediated mechanosensing and abnormal angiogenesis in vitro. Proc Natl Acad Sci USA 2008; 105(32): 11305-10.
[http://dx.doi.org/10.1073/pnas.0800835105] [PMID: 18685096]

[86] Krishnan R, Klumpers DD, Park CY, *et al.* Substrate stiffening promotes endothelial monolayer disruption through enhanced physical forces. Am J Physiol Cell Physiol 2011; 300(1): C146-54.
[http://dx.doi.org/10.1152/ajpcell.00195.2010] [PMID: 20861463]

[87] Huynh J, Nishimura N, Rana K, *et al.* Age-related intimal stiffening enhances endothelial permeability and leukocyte transmigration. Sci Transl Med 2011; 3(112): 112ra122.
[http://dx.doi.org/10.1126/scitranslmed.3002761] [PMID: 22158860]

[88]   Birukova AA, Tian X, Tian Y, Higginbotham K, Birukov KG. Rap-afadin axis in control of Rho signaling and endothelial barrier recovery. Mol Biol Cell 2013; 24(17): 2678-88.
[http://dx.doi.org/10.1091/mbc.e13-02-0098] [PMID: 23864716]

[89]   Bayless KJ, Davis GE. Microtubule depolymerization rapidly collapses capillary tube networks in vitro and angiogenic vessels *in vivo* through the small GTPase Rho. J Biol Chem 2004; 279(12): 11686-95.
[http://dx.doi.org/10.1074/jbc.M308373200] [PMID: 14699132]

[90]   Wei WC, Lin HH, Shen MR, Tang MJ. Mechanosensing machinery for cells under low substratum rigidity. Am J Physiol Cell Physiol 2008; 295(6): C1579-89.
[http://dx.doi.org/10.1152/ajpcell.00223.2008] [PMID: 18923058]

[91]   Craig DH, Haimovich B, Basson MD. Alpha-actinin-1 phosphorylation modulates pressure-induced colon cancer cell adhesion through regulation of focal adhesion kinase-Src interaction. Am J Physiol Cell Physiol 2007; 293(6): C1862-74.
[http://dx.doi.org/10.1152/ajpcell.00118.2007] [PMID: 17898132]

[92]   Byfield FJ, Reen RK, Shentu TP, Levitan I, Gooch KJ. Endothelial actin and cell stiffness is modulated by substrate stiffness in 2D and 3D. J Biomech 2009; 42(8): 1114-9.
[http://dx.doi.org/10.1016/j.jbiomech.2009.02.012] [PMID: 19356760]

[93]   Shen CJ, Raghavan S, Xu Z, *et al.* Decreased cell adhesion promotes angiogenesis in a Pyk2-dependent manner. Exp Cell Res 2011; 317(13): 1860-71.
[http://dx.doi.org/10.1016/j.yexcr.2011.05.006] [PMID: 21640103]

[94]   Fischer RS, Gardel M, Ma X, Adelstein RS, Waterman CM. Local cortical tension by myosin II guides 3D endothelial cell branching. Curr Biol 2009; 19(3): 260-5.
[http://dx.doi.org/10.1016/j.cub.2008.12.045] [PMID: 19185493]

[95]   Hutchings H, Ortega N, Plouët J. Extracellular matrix-bound vascular endothelial growth factor promotes endothelial cell adhesion, migration, and survival through integrin ligation. FASEB J 2003; 17(11): 1520-2.
[http://dx.doi.org/10.1096/fj.02-0691fje] [PMID: 12709411]

[96]   Li J, Zhang YP, Kirsner RS. Angiogenesis in wound repair: angiogenic growth factors and the extracellular matrix. Microsc Res Tech 2003; 60(1): 107-14.
[http://dx.doi.org/10.1002/jemt.10249] [PMID: 12500267]

[97]   Wu Y, Al-Ameen MA, Ghosh G. Integrated effects of matrix mechanics and vascular endothelial growth factor (VEGF) on capillary sprouting. Ann Biomed Eng 2014; 42(5): 1024-36.
[http://dx.doi.org/10.1007/s10439-014-0987-7] [PMID: 24558074]

[98]   Santos L, Fuhrmann G, Juenet M, *et al.* Extracellular stiffness modulates the expression of functional proteins and growth factors in endothelial cells. Adv Healthc Mater 2015; 4(14): 2056-63.
[http://dx.doi.org/10.1002/adhm.201500338] [PMID: 26270789]

[99]   Krishnan L, Hoying JB, Nguyen H, Song H, Weiss JA. Interaction of angiogenic microvessels with the extracellular matrix. Am J Physiol Heart Circ Physiol 2007; 293(6): H3650-8.
[http://dx.doi.org/10.1152/ajpheart.00772.2007] [PMID: 17933969]

[100]   Oberleithner H, Riethmüller C, Schillers H, MacGregor GA, de Wardener HE, Hausberg M. Plasma sodium stiffens vascular endothelium and reduces nitric oxide release. Proc Natl Acad Sci USA 2007; 104(41): 16281-6.
[http://dx.doi.org/10.1073/pnas.0707791104] [PMID: 17911245]

[101]   Oberleithner H, Callies C, Kusche-Vihrog K, *et al.* Potassium softens vascular endothelium and increases nitric oxide release. Proc Natl Acad Sci USA 2009; 106(8): 2829-34.
[http://dx.doi.org/10.1073/pnas.0813069106] [PMID: 19202069]

[102]   Szczygiel AM, Brzezinka G, Targosz-Korecka M, Chlopicki S, Szymonski M. Elasticity changes anti-correlate with NO production for human endothelial cells stimulated with TNF-α. Pflugers Arch 2012;

463(3): 487-96.
[http://dx.doi.org/10.1007/s00424-011-1051-1] [PMID: 22160395]

[103] Kondrikov D, Fonseca FV, Elms S, *et al.* Beta-actin association with endothelial nitric-oxide synthase modulates nitric oxide and superoxide generation from the enzyme. J Biol Chem 2010; 285(7): 4319-27.
[http://dx.doi.org/10.1074/jbc.M109.063172] [PMID: 19946124]

[104] Fels J, Callies C, Kusche-Vihrog K, Oberleithner H. Nitric oxide release follows endothelial nanomechanics and not vice versa. Pflugers Arch 2010; 460(5): 915-23.
[http://dx.doi.org/10.1007/s00424-010-0871-8] [PMID: 20809399]

[105] Boopathy GTK, Kulkarni M, Ho SY, *et al.* Cavin-2 regulates the activity and stability of endothelial nitric-oxide synthase (eNOS) in angiogenesis. J Biol Chem 2017; 292(43): 17760-76.
[http://dx.doi.org/10.1074/jbc.M117.794743] [PMID: 28912276]

[106] Coletta C, Papapetropoulos A, Erdelyi K, *et al.* Hydrogen sulfide and nitric oxide are mutually dependent in the regulation of angiogenesis and endothelium-dependent vasorelaxation. Proc Natl Acad Sci USA 2012; 109(23): 9161-6.
[http://dx.doi.org/10.1073/pnas.1202916109] [PMID: 22570497]

[107] Lesman A, Rosenfeld D, Landau S, Levenberg S. Mechanical regulation of vascular network formation in engineered matrices. Adv Drug Deliv Rev 2016; 96: 176-82.
[http://dx.doi.org/10.1016/j.addr.2015.07.005] [PMID: 26212159]

[108] Jalali S, Li YS, Sotoudeh M, *et al.* Shear stress activates p60src-Ras-MAPK signaling pathways in vascular endothelial cells. Arterioscler Thromb Vasc Biol 1998; 18(2): 227-34.
[http://dx.doi.org/10.1161/01.ATV.18.2.227] [PMID: 9484987]

[109] Shikata Y, Rios A, Kawkitinarong K, DePaola N, Garcia JG, Birukov KG. Differential effects of shear stress and cyclic stretch on focal adhesion remodeling, site-specific FAK phosphorylation, and small GTPases in human lung endothelial cells. Exp Cell Res 2005; 304(1): 40-9.
[http://dx.doi.org/10.1016/j.yexcr.2004.11.001] [PMID: 15707572]

[110] Chen Z, Tzima E. PECAM-1 is necessary for flow-induced vascular remodeling. Arterioscler Thromb Vasc Biol 2009; 29(7): 1067-73.
[http://dx.doi.org/10.1161/ATVBAHA.109.186692] [PMID: 19390054]

[111] Ozaki CK, Jiang Z, Berceli SA. TNF-alpha and shear stress-induced large artery adaptations. J Surg Res 2007; 141(2): 299-305.
[http://dx.doi.org/10.1016/j.jss.2006.12.563] [PMID: 17574273]

[112] Singh TM, Abe KY, Sasaki T, Zhuang YJ, Masuda H, Zarins CK. Basic fibroblast growth factor expression precedes flow-induced arterial enlargement. J Surg Res 1998; 77(2): 165-73.
[http://dx.doi.org/10.1006/jsre.1998.5376] [PMID: 9733604]

[113] Yu J, Bergaya S, Murata T, *et al.* Direct evidence for the role of caveolin-1 and caveolae in mechanotransduction and remodeling of blood vessels. J Clin Invest 2006; 116(5): 1284-91.
[http://dx.doi.org/10.1172/JCI27100] [PMID: 16670769]

[114] Zhang H, Sunnarborg SW, McNaughton KK, Johns TG, Lee DC, Faber JE. Heparin-binding epidermal growth factor-like growth factor signaling in flow-induced arterial remodeling. Circ Res 2008; 102(10): 1275-85.
[http://dx.doi.org/10.1161/CIRCRESAHA.108.171728] [PMID: 18436796]

[115] Helisch A, Schaper W. Arteriogenesis: the development and growth of collateral arteries. Microcirculation 2003; 10(1): 83-97.
[http://dx.doi.org/10.1080/mic.10.1.83.97] [PMID: 12610665]

[116] Lamalice L, Le Boeuf F, Huot J. Endothelial cell migration during angiogenesis. Circ Res 2007; 100(6): 782-94.
[http://dx.doi.org/10.1161/01.RES.0000259593.07661.1e] [PMID: 17395884]

[117] Yamamoto K, Takahashi T, Asahara T, *et al.* Proliferation, differentiation, and tube formation by endothelial progenitor cells in response to shear stress. J Appl Physiol 2003; 95(5): 2081-8.
[http://dx.doi.org/10.1152/japplphysiol.00232.2003] [PMID: 12857765]

[118] Ito WD, Arras M, Scholz D, Winkler B, Htun P, Schaper W. Angiogenesis but not collateral growth is associated with ischemia after femoral artery occlusion. Am J Physiol 1997; 273(3 Pt 2): H1255-65.
[PMID: 9321814]

[119] Kubis N, Checoury A, Tedgui A, Lévy BI. Adaptive common carotid arteries remodeling after unilateral internal carotid artery occlusion in adult patients. Cardiovasc Res 2001; 50(3): 597-602.
[http://dx.doi.org/10.1016/S0008-6363(01)00206-1] [PMID: 11376636]

[120] Baumbach GL, Ghoneim S. Vascular remodeling in hypertension. Scanning Microsc 1993; 7(1): 137-42.
[PMID: 8316787]

[121] Hoefer IE, den Adel B, Daemen MJ. Biomechanical factors as triggers of vascular growth. Cardiovasc Res 2013; 99(2): 276-83.
[http://dx.doi.org/10.1093/cvr/cvt089] [PMID: 23580605]

[122] Moroni L, Schotel R, Sohier J, de Wijn JR, van Blitterswijk CA. Polymer hollow fiber three-dimensional matrices with controllable cavity and shell thickness. Biomaterials 2006; 27(35): 5918-26.
[http://dx.doi.org/10.1016/j.biomaterials.2006.08.015] [PMID: 16935328]

[123] Luo Y, Lode A, Gelinsky M. Direct plotting of three-dimensional hollow fiber scaffolds based on concentrated alginate pastes for tissue engineering. Adv Healthc Mater 2013; 2(6): 777-83.
[http://dx.doi.org/10.1002/adhm.201200303] [PMID: 23184455]

[124] Sun B, Jiang XJ, Zhang S, *et al.* Electrospun anisotropic architectures and porous structures for tissue engineering. J Mater Chem B Mater Biol Med 2015; 3(27): 5389-410.
[http://dx.doi.org/10.1039/C5TB00472A] [PMID: 32262511]

[125] Malinauskas M, Rekštytė S, Lukoševičius L, *et al.* 3D Microporous Scaffolds Manufactured *via* Combination of Fused Filament Fabrication and Direct Laser Writing Ablation. Micromachines (Basel) 2014; 5: 839-58.
[http://dx.doi.org/10.3390/mi5040839]

[126] Borenstein JT, Terai H, King KR, Weinberg EJ, Kaazempur-Mofrad MR, Vacanti JP. Microfabrication technology for vascularized tissue engineering. Biomed Microdevices 2002; 4: 167-75.
[http://dx.doi.org/10.1023/A:1016040212127]

[127] Ye X, Lu L, Kolewe ME, *et al.* A biodegradable microvessel scaffold as a framework to enable vascular support of engineered tissues. Biomaterials 2013; 34(38): 10007-15.
[http://dx.doi.org/10.1016/j.biomaterials.2013.09.039] [PMID: 24079890]

[128] Miller JS, Stevens KR, Yang MT, *et al.* Rapid casting of patterned vascular networks for perfusable engineered three-dimensional tissues. Nat Mater 2012; 11(9): 768-74.
[http://dx.doi.org/10.1038/nmat3357] [PMID: 22751181]

[129] Zheng Y, Chen J, Craven M, *et al.* In vitro microvessels for the study of angiogenesis and thrombosis. Proc Natl Acad Sci USA 2012; 109(24): 9342-7.
[http://dx.doi.org/10.1073/pnas.1201240109] [PMID: 22645376]

[130] Zhao L, Lee VK, Yoo SS, Dai G, Intes X. The integration of 3-D cell printing and mesoscopic fluorescence molecular tomography of vascular constructs within thick hydrogel scaffolds. Biomaterials 2012; 33(21): 5325-32.
[http://dx.doi.org/10.1016/j.biomaterials.2012.04.004] [PMID: 22531221]

[131] Nichol JW, Koshy ST, Bae H, Hwang CM, Yamanlar S, Khademhosseini A. Cell-laden microengineered gelatin methacrylate hydrogels. Biomaterials 2010; 31(21): 5536-44.
[http://dx.doi.org/10.1016/j.biomaterials.2010.03.064] [PMID: 20417964]

[132] Sadr N, Zhu M, Osaki T, *et al.* SAM-based cell transfer to photopatterned hydrogels for microengineering vascular-like structures. Biomaterials 2011; 32(30): 7479-90.
[http://dx.doi.org/10.1016/j.biomaterials.2011.06.034] [PMID: 21802723]

[133] Linville RM, Boland NF, Covarrubias G, Price GM, Tien J. Physical and chemical signals that promote vascularization of capillary-scale channels. Cell Mol Bioeng 2016; 9(1): 73-84.
[http://dx.doi.org/10.1007/s12195-016-0429-8] [PMID: 27110295]

[134] Tsuda Y, Shimizu T, Yamato M, *et al.* Cellular control of tissue architectures using a three-dimensional tissue fabrication technique. Biomaterials 2007; 28(33): 4939-46.
[http://dx.doi.org/10.1016/j.biomaterials.2007.08.002] [PMID: 17709135]

[135] Nikkhah M, Eshak N, Zorlutuna P, *et al.* Directed endothelial cell morphogenesis in micropatterned gelatin methacrylate hydrogels. Biomaterials 2012; 33(35): 9009-18.
[http://dx.doi.org/10.1016/j.biomaterials.2012.08.068] [PMID: 23018132]

[136] Hoch E, Tovar GEM, Borchers K. Bioprinting of artificial blood vessels: current approaches towards a demanding goal. Eur J Cardiothorac Surg 2014; 46(5): 767-778.
[http://dx.doi.org/10.1093/ejcts/ezu242]

[137] Ozbolat IT, Yu Y. Bioprinting toward organ fabrication: challenges and future trends. IEEE Trans Biomed Eng 2013; 60(3): 691-9.
[http://dx.doi.org/10.1109/TBME.2013.2243912] [PMID: 23372076]

[138] Lu J, Rao MP, MacDonald NC, Khang D, Webster TJ. Improved endothelial cell adhesion and proliferation on patterned titanium surfaces with rationally designed, micrometer to nanometer features. Acta Biomater 2008; 4(1): 192-201.
[http://dx.doi.org/10.1016/j.actbio.2007.07.008] [PMID: 17851147]

[139] Tsai SH, Liu YW, Tang WC, *et al.* Characterization of porcine arterial endothelial cells cultured on amniotic membrane, a potential matrix for vascular tissue engineering. Biochem Biophys Res Commun 2007; 357(4): 984-90.
[http://dx.doi.org/10.1016/j.bbrc.2007.04.047] [PMID: 17459341]

[140] Tajima S, Chu JS, Li S, Komvopoulos K. Differential regulation of endothelial cell adhesion, spreading, and cytoskeleton on low-density polyethylene by nanotopography and surface chemistry modification induced by argon plasma treatment. J Biomed Mater Res A 2008; 84(3): 828-36.
[http://dx.doi.org/10.1002/jbm.a.31539] [PMID: 17685408]

[141] Chung TW, Liu DZ, Wang SY, Wang SS. Enhancement of the growth of human endothelial cells by surface roughness at nanometer scale. Biomaterials 2003; 24(25): 4655-61.
[http://dx.doi.org/10.1016/S0142-9612(03)00361-2] [PMID: 12951008]

[142] Foley JD, Grunwald EW, Nealey PF, Murphy CJ. Cooperative modulation of neuritogenesis by PC12 cells by topography and nerve growth factor. Biomaterials 2005; 26(17): 3639-44.
[http://dx.doi.org/10.1016/j.biomaterials.2004.09.048] [PMID: 15621254]

[143] Karuri NW, Liliensiek S, Teixeira AI, *et al.* Biological length scale topography enhances cell-substratum adhesion of human corneal epithelial cells. J Cell Sci 2004; 117(Pt 15): 3153-64.
[http://dx.doi.org/10.1242/jcs.01146] [PMID: 15226393]

[144] Samaroo HD, Lu J, Webster TJ. Enhanced endothelial cell density on NiTi surfaces with sub-micron to nanometer roughness. Int J Nanomedicine 2008; 3(1): 75-82.
[PMID: 18488418]

[145] Khang D, Lu J, Yao C, Haberstroh KM, Webster TJ. The role of nanometer and sub-micron surface features on vascular and bone cell adhesion on titanium. Biomaterials 2008; 29(8): 970-83.
[http://dx.doi.org/10.1016/j.biomaterials.2007.11.009] [PMID: 18096222]

[146] Flemming RG, Murphy CJ, Abrams GA, Goodman SL, Nealey PF. Effects of synthetic micro- and nano-structured surfaces on cell behavior. Biomaterials 1999; 20(6): 573-88.
[http://dx.doi.org/10.1016/S0142-9612(98)00209-9] [PMID: 10213360]

[147] Karuri NW, Porri TJ, Albrecht RM, Murphy CJ, Nealey PF. Nano- and microscale holes modulate cell-substrate adhesion, cytoskeletal organization, and -beta1 integrin localization in SV40 human corneal epithelial cells. IEEE Trans Nanobioscience 2006; 5(4): 273-80.
[http://dx.doi.org/10.1109/TNB.2006.886570] [PMID: 17181027]

[148] Teixeira AI, Nealey PF, Murphy CJ. Responses of human keratocytes to micro- and nanostructured substrates. J Biomed Mater Res A 2004; 71(3): 369-76.
[http://dx.doi.org/10.1002/jbm.a.30089] [PMID: 15470741]

[149] Bettinger CJ, Langer R, Borenstein JT. Engineering substrate topography at the micro- and nanoscale to control cell function. Angew Chem Int Ed Engl 2009; 48(30): 5406-15.
[http://dx.doi.org/10.1002/anie.200805179] [PMID: 19492373]

[150] Cavalcanti-Adam EA, Aydin D, Hirschfeld-Warneken VC, Spatz JP. Cell adhesion and response to synthetic nanopatterned environments by steering receptor clustering and spatial location. HFSP J 2008; 2(5): 276-85.
[http://dx.doi.org/10.2976/1.2976662] [PMID: 19404439]

[151] Chai C, Leong KW. Biomaterials approach to expand and direct differentiation of stem cells. Mol Ther 2007; 15(3): 467-80.
[http://dx.doi.org/10.1038/sj.mt.6300084] [PMID: 17264853]

[152] Yim EK, Pang SW, Leong KW. Synthetic nanostructures inducing differentiation of human mesenchymal stem cells into neuronal lineage. Exp Cell Res 2007; 313(9): 1820-9.
[http://dx.doi.org/10.1016/j.yexcr.2007.02.031] [PMID: 17428465]

[153] Bettinger CJ, Zhang Z, Gerecht S, Borenstein JT, Langer R. Enhancement of *in vitro* capillary tube formation by substrate nanotopography. Adv Mater 2008; 20(1): 99-103.
[http://dx.doi.org/10.1002/adma.200702487] [PMID: 19440248]

[154] Chaturvedi RR, Stevens KR, Solorzano RD, *et al.* Patterning vascular networks *in vivo* for tissue engineering applications. Tissue Eng Part C Methods 2015; 21(5): 509-17.
[http://dx.doi.org/10.1089/ten.tec.2014.0258] [PMID: 25390971]

[155] BP T. Mechanobiology of adult and stem cells. International review of cells & molecular. Biology (Basel) 2008; 271: 301-46.

[156] Hayashi K, Naiki T. Adaptation and remodeling of vascular wall; biomechanical response to hypertension. J Mech Behav Biomed Mater 2009; 2(1): 3-19.
[http://dx.doi.org/10.1016/j.jmbbm.2008.05.002] [PMID: 19627803]

[157] Yano Y, Geibel J, Sumpio BE. Tyrosine phosphorylation of pp125FAK and paxillin in aortic endothelial cells induced by mechanical strain. Am J Physiol 1996; 271(2 Pt 1): C635-49.
[http://dx.doi.org/10.1152/ajpcell.1996.271.2.C635] [PMID: 8770005]

[158] Kim S, Lee H, Chung M, Jeon NL. Engineering of functional, perfusable 3D microvascular networks on a chip. Lab Chip 2013; 13(8): 1489-500.
[http://dx.doi.org/10.1039/c3lc41320a] [PMID: 23440068]

[159] Donald AE, Halcox JP, Charakida M, *et al.* Methodological approaches to optimize reproducibility and power in clinical studies of flow-mediated dilation. J Am Coll Cardiol 2008; 51(20): 1959-64.
[http://dx.doi.org/10.1016/j.jacc.2008.02.044] [PMID: 18482664]

[160] Mc Loughlin S, Mc Loughlin MJ, Pyke KE, Padilla J, Atkinson G, Harris RA, *et al.* Letter to the editor: "Assessment of flow-mediated dilation in humans: a methodological and physiological guideline". Am J Physiol Heart Circ Physiol 2011; 300(2): H712.
[http://dx.doi.org/10.1152/ajpheart.01143.2010] [PMID: 21282474]

[161] Davies PF. Hemodynamic shear stress and the endothelium in cardiovascular pathophysiology. Nat Clin Pract Cardiovasc Med 2009; 6(1): 16-26.
[http://dx.doi.org/10.1038/ncpcardio1397] [PMID: 19029993]

[162] Corson MA, James NL, Latta SE, Nerem RM, Berk BC, Harrison DG. Phosphorylation of endothelial nitric oxide synthase in response to fluid shear stress. Circ Res 1996; 79(5): 984-91.
[http://dx.doi.org/10.1161/01.RES.79.5.984] [PMID: 8888690]

CHAPTER 3

Microenvironment of Pathological Vascularization

Mei Zhang and **Xiaoxiao Cai**[*]

State Key Laboratory of Oral Diseases, West China Hospital of Stomatology, Sichuan University, Chengdu 610041, China

Abstract: Past studies have shown that many destructive diseases drive abnormal angiogenesis and progression, such as inflammatory diseases, cancers, rheumatoid arthritis, and atherosclerosis diseases. These diseases, which have a variety of consequences, show some common pathophysiological characteristics, among which the proliferation of endothelial cells, recruitment of immune cells, and high-expression of angiogenic factors play a key role. At the same time, local hypoxia, inflammation, senile, and local ischaemia cause adverse consequences such as abnormal vascularisation. Abnormal blood vessels usually include vascular structural abnormalities, abnormal endothelial cells, excessive vascular permeability, vascular dysfunction, *etc*. The pathological microenvironment is related to abnormal vascularisation and further aggravates the abnormality of vascularisation. Therefore, this review will be helpful for further study of vascularised tissue engineering.

Keywords: Angiogenesis, Ageing, Inflammation, Ischemia and Hypoxia, Pathological Vascularization.

1. INTRODUCTION

Angiogenesis, the formation of new blood vessels from pre-existing vessels, is necessary for various physiological processes. During evolution, blood vessels emerge, and transport oxygen and nutrients to the distant tissues and organs [1]. Undoubtedly, these blood vessels are essential for the growth of tissues in embryos and repair of injured tissues in adults. However, the imbalance of vascular growth leads to many diseases. Similarly, the pathological microenvironment of these diseases also accelerates the formation of abnormal blood vessels [2, 3]. As a matter of fact, pathological angiogenesis is a marker of malignant tumours, and various metastatic, ischaemic, and inflammatory diseases [4]. The pathological microenvironment of vessels is a dynamic network, which is composed of many factors.

---

[*] **Corresponding author Xiaoxiao Cai:** Sichuan University, West China School of Stomatology, China; E-mail: xcai@scu.edu.cn

Microenvironment has an increasingly significant effect on angiogenesis under the conditions of diseases. On the other hand, angiogenesis is a marker of growth, invasion, and metastasis of malignant diseases. Increasing studies have shown that the microenvironment of diseases is closely related to pathological angiogenesis. In this review, the effects of the microenvironment of tumours, and various metastatic, ischaemic, and inflammatory diseases on angiogenesis are analysed in detail to provide a broader idea for angiogenesis research, especially for understanding angiogenesis in tissue engineering.

2.   PATHOLOGICAL   CONDITIONS   LEADING   TO   ADVERSE VASCULARISATION

Pathological angiogenesis is a sign of cancers and various ischaemic and inflammatory diseases [4], as shown in Fig. (**1**). By exploring the formation of abnormal blood vessels, their mode of action, the key molecules, and microenvironments involved, and related studies, it would help us understand the occurrence and development of cancers, and ischaemic, and inflammatory diseases [5, 6]. During evolution, blood vessels emerge and then transport oxygen and nutrients to distant tissues [7, 8]. Undeniably, these blood vessels are essential for sustaining the growth of normal and injured tissues. However, abnormal, excessive, and insufficient vascular growth is usually involved in the pathogenesis of many diseases [2]. After birth, angiogenesis continues to play a significant role in organ growth. While the growth of blood vessels is stationary in adulthood, angiogenesis occurs only during pregnancy [1]. Nevertheless, endothelial cells manifcst thcir rcmarkablc divisivc capacity in thc facc of pathological stimulation, such as vascular hypoxia and lymphangitis [9]. During healing and repair of wound, angiogenesis is reactivated. However, in several other diseases, excessive stimulus can break out the balance between stimulants and inhibitors, which lead to abnormal angiogenesis [10]. The most common conditions for activation of abnormal angiogenesis are inflammatory diseases and malignant tumours, however, other diseases can also have an impact, such as asthma, obesity, cirrhosis of the liver, diabetes, endometriosis, AIDS, multiple sclerosis, autoimmune diseases, and bacterial infections [1, 11].

2.1. Inflammation-Related Diseases and Angiogenesis

Angiogenesis participate in the occurrence and progression of many destructive diseases, such as cancers, atherosclerosis, inflammatory bowel disease [IBD], and rheumatoid arthritis [12]. Angiogenesis is regulated by different types of cells in its microenvironment, such as immune cells [13 - 16]. During inflammation, both endothelial and parietal cells are involved in the migration of white blood cells to inflammatory centres [17, 18]. Immune cells participate in wound healing by controlling the formation of new blood vessels. Emerging research has shown that these cells are involved in promoting angiogenesis and vascular remodelling in ischaemic injury, cancers, and wound healing [19, 20]. These immune cells promote angiogenesis under the strong influence of their microenvironment, especially the inflammatory microenvironment [21].

2.1.1. Inflammation-Related Diseases and Angiogenesis

Moderate inflammation is important for maintaining homeostasis [22], but severe and persistent inflammation leads to some chronic inflammatory diseases, such as cancers [21], atherosclerosis [23], IBD, and rheumatoid arthritis [24]. Hypoxia-inducible factors [HIFs] are produced by inflammatory tissues, which are finished by activating endothelial cells, macrophages, and fibroblasts due to local hypoxia [25 - 27]. Long-term infiltration of these factors leads to adhesive degradation of endothelial cells and microvascular instability [28, 29]. On the other hand, it is beneficial to the proliferation and migration of endothelial cells and angiogenesis [30, 31].

Although these diseases develop differently and have different clinical manifestations, they have the same pathophysiological peculiarity, among which enhanced vascular permeability, macrophage polarization, and monocyte recruitment play essential roles [12]. Tumour-associated macrophages [TAM] excrete some angiogenic growth factors and proteases, which include vascular endothelial growth factor [VEGF], to promote angiogenesis and further migration and infiltration of tumours [32 - 34]. Atherosclerotic plaque can be formed by some proteolytic enzymes, which are secreted by macrophages, including matrix metalloproteinases [MMPs] [35]. Similar to angiogenesis of tumours, neovascularisation is mediated by macrophage-derived angiogenic factors in atherosclerotic plaques [36]. Other inflammatory diseases function similarly. Local hypoxia and increased vascular permeability caused by inflammation lead to new angiogenesis. During this period, the related pro-inflammatory factors and angiogenic factors secreted by macrophages play a key role.

2.1.2. Pathological Mechanism of Inflammatory Angiogenesis

First, aggregation of immune cells plays an important role in the process of inflammatory angiogenesis. Immunocytes associated with angiogenesis include natural killer cells, neutrophils, dendritic cells, eosinophils, mast cells, macrophages, and T cells [37]. It has been reported that angiogenesis is induced by macrophages in wounds and tumours, which stimulate ischaemia-induced arteriogenesis, and promote vascular anastomosis [20, 38 - 40]. Macrophages can secrete several known factors affecting new angiogenesis, including insulin-like growth factor-1 [IGF-1], transforming growth factor-beta [TGF-beta], basic fibroblast growth factor [bFGF], VEGF, interleukin 8 [IL-8] and tumour necrosis factor-alpha [TNF-alpha] [41]. Dendritic cells are known to regulate angiogenesis in tumours [37]. By producing TNF-alpha and IL-8, plasmocyte-like dendritic cells [PDC] promote angiogenesis in ovarian cancers [42].

Granulocytes, including mast cells, neutrophils, and eosinophils, are thought to be involved in the early stage of angiogenesis [37]. In ischaemic tissues, granulocytes affect angiogenesis by activating granulocyte-macrophage colony-stimulating factor [GM-CSF] to induce the formation of VEGF [43, 44]. Besides, they excrete matrix metalloproteinase-9 [MMP-9] to promote angiogenesis of tumours [45]. Eosinophils and Mast cells are the effective sources of angiogenic factors such as bFGF, IL-8, GM-CSF, and TNF-a [46, 47]. Lymphocytes, including natural killer cells and T cells, can directly promote angiogenesis by secreting angiogenesis-related factors. Furthermore, lymphocytes can promote angiogenesis through monocytes [48 - 51].

In addition, inflammation is involved in the pathological development of rheumatoid arthritis and IBD, and further consequences lead to increased vascular permeability [52, 53]. Other diseases leading to increased vascular permeability include tumours, wound healing, and chronic inflammatory diseases such as psoriasis and cellular immunity [54, 55]. Due to increased vascular permeability, fibrin and certain tissue factors are widely exuded [56]. Fibrin induces the neovascularisation and inward growth of fibroblasts. Morcover, fibrin protects them from degradation by isolating growth factors, and induces the expression of angiogenic molecules such as IL-8 and tissue factors [29].

2.1.3. Consequence of Inflammatory Angiogenesis

Inflammatory mediators stimulate angiogenesis, in which macrophages play an important role, and most macrophages are found in these sites with abnormal angiogenesis [57]. However, abnormal angiogenesis often has adverse consequences; the structure of blood vessels tends to be immature [58]. With the increase of vascular density, the infiltration of macrophages and the expression of VEGF in synovium also increased. Next, the proliferation of endothelial cells also increased to form more vessels [59]. Increased microvessel density and the expression of angiogenic factors are associated with further changes in hypoxia-inducible factor-1alpha [HIF-1a], suggesting that hypoxia also plays a significant part in the duration of angiogenesis [60]. The newly formed blood vessels have higher permeability and are easy to form edema [61]. Adhesion factors are highly expressed in neovascularisation, which can easily lead to inflammatory cell infiltration [62, 63]. Neovascularisation provides oxygen and nutrients for further development of inflammation [30]. In conclusion, inflammation and blood vessels influence each other, that is, inflammation leads to angiogenesis, which in turn accelerates the inflammatory reaction.

2.2. Vascularization Under Ischemia

In the case of ischaemia, the vessel chamber undergoes complex molecular and cellular changes, which determine the extent of perfusion recovery in ischaemic tissues. In ischaemic diseases, the harmful microenvironment associated with tissues of ischaemia significantly weakens most of the pathways involved in vascular growth and vasculogenesis [64]. In addition, cardiovascular risk factors also participate in the formation of inhibiting environment, which inhibits the revascularisation after ischaemia [65]. Tissue repair and remodelling of acute and chronic ischaemic vascular diseases mainly involve angiogenesis, vasculogenesis, collateral growth, and arteriogenesis, and these processes complement and influence each other [19].

2.2.1. Ischemic Diseases and Angiogenesis

Ischaemic diseases constitute a group of cardiovascular diseases caused by insufficient oxygen supply to tissues [such as the heart], the brain [cerebrovascular disease], and surrounding muscles [peripheral arterial disease] [66]. Atherosclerosis is the main pathophysiological process leading to ischaemic diseases. Atherosclerosis, thrombosis, and vascular occlusion can lead to further ischaemia of the heart, also known as myocardial infarction [67]. Thrombosis is also a common cause of cerebral ischaemic injury. Ischaemia limits the use of oxygen and nutrients. In particular, ischaemia-reperfusion injury results in a large number of reactive oxygen species [ROS], which impairs the function of endothelial cells, leads to endothelial cell apoptosis, and affects angiogenesis [19] Atherosclerosis and subsequent arterial thrombosis can lead to acute and life-threatening cardiovascular events, such as myocardial infarction and stroke [66].

2.2.2. Pathological Mechanism of Ischemic Angiogenesis

Ischaemia-induced deprivation of glucose and oxygen can lead to the activation of VEGF [68]. Experiments have shown that over-stress does not induce the increase of VEGF in cells with impaired metabolism, and that the microenvironment of hypoxia and deficiency of glucose and other nutrients have a certain effect on VEGF [68]. The level of VEGF is regulated by hypoxia and hypoglycemia [68 -

70]. The transcriptional regulation of VEGF is mediated by HIF-1, which accumulates due to the stabilisation of highly unstable proteins under hypoxia [71, 72]. It is not clear whether hypoxia-induced transcription factor endothelial PAS domain protein 1[EPAS-1] also regulates the expression of VEGF. EPAS-1 may play a role in the crosstalk between endothelial cells and smooth muscle cells or pericytes, because it induces the expression of tie-2 [73], which is a specific receptor of endothelial cells [74], and ligand angiopoietin-1 may also play a role in the process of vascular remodelling [75, 76].

2.2.3. Consequence of Ischemic Angiogenesis

Ischaemia-driven angiogenesis is mainly an adaptive physiological response to increased tissue mass or oxygen consumption. Usually, new angiogenesis manifests an increase in vascular density [77]. Ischaemia-induced angiogenesis should not be excessive, since it might cause serious pathological changes. For example, different forms of retinopathy are caused by excessive vascular growth. After vascular occlusion, compensatory angiogenesis is performed for severe retinal ischaemia. However, if the blood vessels grow excessively, they will leak abnormally. The blood vessels in the retina may grow into the vitreous body and cause retinal detachment [78, 79]. Local ischaemia caused by vascular occlusion is usually beneficial to the formation of collateral vessels, including microvascular and macrovascular collateral circulations.

In addition, ischaemia can also be caused by vascular damage resulting from wounds. In fact, ischaemia-mediated angiogenesis is a major factor in neovascularisation associated with wound healing [80]. However, during hypoxia, capillary formation stops beyond a certain extent. At the same time, stress can also cause angiogenesis. It has been reported that stress-induced angiogenesis is an important component of neovascularisation in tumours.

2.3. Vascularization Under Hypoxia

Hypoxia plays a regulatory role in angiogenesis. It is an important part of the mechanism to maintain homeostasis by linking vascular oxygen supply with metabolic demand [81]. Previous studies have shown that many tumours have a severe hypoxic microenvironment, and promote angiogenesis by secreting some angiogenesis stimulating factors [82, 83]. Trauma can also cause hypoxic conditions [84]. In wound repair, hypoxia induces bone marrow-derived macrophages to secrete angiogenic factors to promote angiogenesis [85]. In tumours, the VEGF has upregulated expression in the surrounding necrotic areas, which suggests that hypoxia may be a mechanism for stimulating angiogenesis in tumours [86, 87]. In conclusion, these studies suggest that oxygen may be an important regulator of angiogenesis, and that hypoxia caused by oxygen itself or energy metabolism may be the trigger point of some consequences [88].

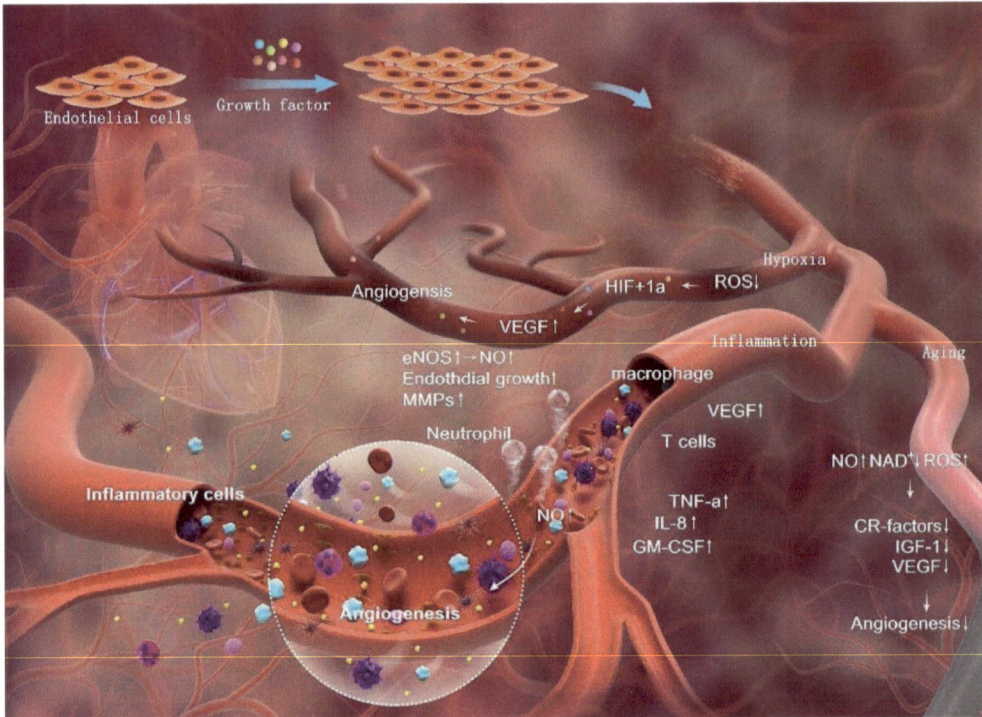

Fig. (1). Angiogenesis is achieved by proliferation and migration of endothelial cells under the influence of related angiogenic factors. Some pathological angiogenesis microenvironment, mainly including ischemia, hypoxia, inflammation and aging, play a significant role in angiogenesis. The angiogenesis of the blood vessel is influenced by the related mechanisms and some related the growth factor and cells. These blood vessels are mass generated or reduced, and the local focus is transmitted to the remote through the new blood vessel and the blood circulation.

2.3.1. Hypoxic Disease and Angiogenesis

Hypoxia is a decrease in oxygen tension in normal tissues, and occurs in lung diseases, acute and chronic vascular diseases, cancers, and ischaemic diseases. Responding to hypoxia, cells show a series of biological reactions, which include activation of signal transduction pathways to adjust angiogenesis, proliferation, and death [89]. Especially, local oxygen deficiency often leads to new angiogenesis to meet physiological needs [88]. For example, collateral angiogenesis is needed in ischaemic wound healing, in addition, angiogenesis is often observed in disease environments such as cancers and atherosclerosis [85, 90]. In conclusion, by promoting angiogenesis, hypoxia may contribute significantly to disease progression and maintenance [91].

2.3.2. Pathological Mechanism of Hypoxic Angiogenesis

Different cells express a variety of gene products involved in angiogenesis, which mostly respond independently to hypoxia in tissue cultures [92]. Growth factors such as fibroblast growth factors and VEGF, angiopoietins [93 - 96], and nitric oxide synthases participate in the regulation of vascular tone [97], and that of genes related to matrix metabolism, which include collagen prolyl hydroxylase, MMP, and plasminogen activator receptors and plasminogen activator inhibitors [98 - 100]. Hypoxia has both positive and negative impact on angiogenesis, such as its effect on proliferation of endothelial cells [101 - 104].

HIF is the main regulator of oxygen tension homeostasis [105]. Previous studies have shown that pre-eclampsia placenta [106], inflammatory macrophages in joints [107], hypoxic preconditioning of retina, and injured skin indicate certain ischaemia, hypoxia, and inflammation that can induce the upregulation of HIF-1a, which may be related to angiogenesis [108, 109]. Under normal oxygen content, HIF-1a is continuously transcribed and translated, and eventually degraded. However, under hypoxic condition, HIF-1 participates in angiogenesis and metabolism by activating and binding with hypoxic reaction elements [110]. Previous studies have shown that HIF, especially HIF-1a, has an important effect on the regulation of angiogenic genes [111 - 114]. For example, during hypoxia, HIF-1 participates in the upregulation of VEGF [93]. At the same time, HIF-1 downregulates platelet reactive protein [PRP], which results in an angiogenesis-promoting environment [115]. From Fig. (**2**), it shows in detail how HIF-1a signaling pathway affects the process of angiogenesis, especially through VEGF signaling pathway.

2.4. Vascularization Under Senility

Aging is the main pathogenic factor of cardiovascular diseases. Aging results in impaired angiogenesis and endothelial dysfunction, which further increases in the prevalence of cardiovascular diseases in the elderly [116, 117]. At the same time, the elderly suffer from ischaemic diseases and need angiogenesis repair more than young people. Angiogenesis is not only a basic physiological stress response, but also an endogenous repairment after ischaemic injury [117]. Therefore, the effect aging on angiogenesis and endothelial function is helpful to understand and manage cardiovascular diseases.

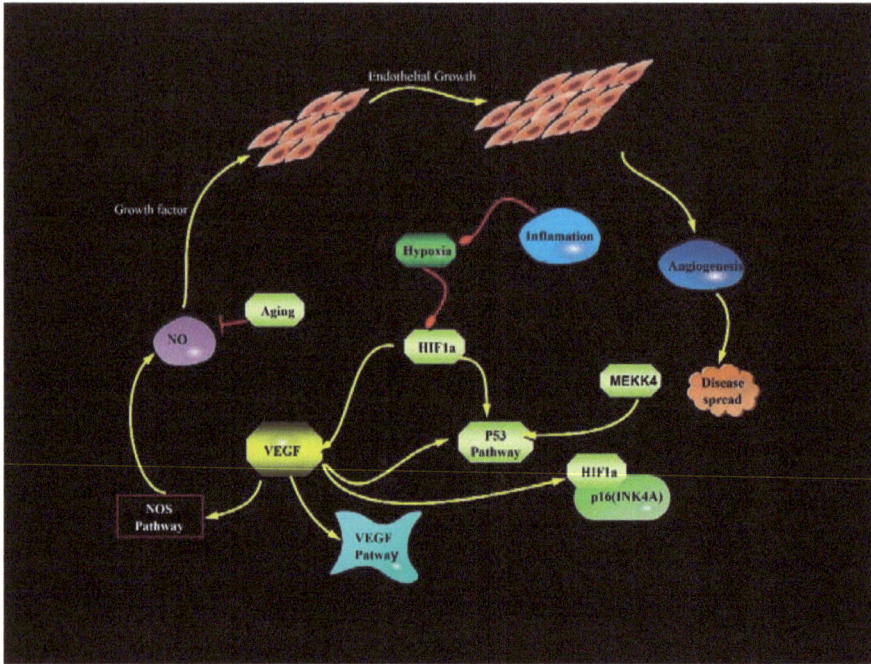

Fig. (2). The effect of pathological microenvironment on angiogenesis through HIF-1-a signaling pathway. In response to a series of stimuli (such as hypoxia, inflammation and aging), tissues respond to the increased expression of secreted angiogenic peptides (such as VEGF) by activating HIF-1-a pathway. The angiogenic peptides regulate endothelial growth and chemotaxis to enhance angiogenesis through eNOS-dependent and –independent mechanisms. Besides, NO itself also regulates these processes, at the same time, it acts on vascular smooth muscle cells to regulate vascular tension, which affects vascular development.

2.4.1. Pathological Mechanism of Aging Angiogenesis

Aging may affect proliferation and the physiological function of cells, especially endothelial cells, and has the greatest influence on angiogenesis [118]. Aging endothelial cells reduce proliferation, restrict the formation of cardiovascular, and easily lead to atherosclerotic plaque formation. Studies have shown that local angiogenesis disorders are associated with senescence of endothelial cells [119]. Telomere shortening may be associated with aging endothelial cells and impaired angiogenesis [120, 121]. In addition to aging, apoptosis of endothelial cells can cause microvascular sparsity by reducing proliferation. The apoptosis of endothelial cells was also related to the increase in senescent cells [122, 123].

Aging-related changes in circulatory factors may also lead to impaired angiogenesis [117]. In particular, changes in angiogenesis-related factors caused by aging are particularly important. These changes include decreased production of VEGF-A and decreased expression of VEGF receptors, which are related to the impaired activation of HIF-1a [124]. Several additional growth factors, which include bFGF and platelet-derived growth factor [PDGF], are impaired in aging endothelial cells [125, 126]. Endocrine changes induced by aging are important contributors to impaired angiogenesis. For example, the reduction in growth hormones and IGF-1, and increase in glucocorticoid cause damage to angiogenesis [127]. Estrogen makes a significant effort in sustaining the function of endothelial cells, stimulating and maintaining the production of nitric oxide, which exists in almost every aspect of angiogenesis [128, 129].

During aging, oxidative stress increases due to the oxidation of DNA, proteins, and lipids [130]. The imbalance between endogenous antioxidant and pro-oxidant molecules, and the increase production of ROS are involved in increasing oxidative stress [131]. The increase in ROS may be related to mitochondrial dysfunction caused by mitochondrial aging [132]. Oxidative stress plays a significant role in angiogenesis. The imbalance in redox will lead to aging, inhibited proliferation, and dysfunction of endothelial cells [133]. Superoxide, including ROS and NO, can affect vasodilation and angiogenesis [117]. In conclusion, the role of aging-related oxidative stress in impaired vascular function and angiogenesis cannot be ignored.

2.4.2. The Consequences of Impaired Angiogenesis

Aging-related impaired angiogenesis results in decreased microvessel density [134, 135] and decreased blood supply, and further deterioration of hypoxic ischaemia in local tissues [134, 136]. Such changes may eventually lead to myocardial ischaemia and infarction, stroke, and peripheral artery diseases [117]. In addition to the aging and apoptosis of endothelial cells, dysfunction of peripheral cells and the downregulation of angiogenic factors lead to microvascular sparsity which is associated with impaired tissue perfusion and the development of hypertension [137].

3. CHARACTERISTICS OF ABNORMAL BLOOD VESSELS

Angiogenesis is an important process of tumour growth and ischaemic diseases. In the normal tissues, angiogenesis promotion and anti-angiogenesis are in a balance, but in pathological condition, such as tumours and acute leukaemia, this balance is broken. Normal blood vessels consist of well-differentiated arteries and veins, small veins, and capillaries, arranged in neat layers and evenly spaced. However, the blood vessels in tumours are different from those in normal tissues; vascular stratification is disordered, and the function and structure are also abnormal [138]. These abnormal vascular structures lead to insufficient nutrient supply and insufficient removal of carbon dioxide and metabolites, which further lead to an acidic microenvironment [139]. Moreover, the combination of these consequences leads to insufficient metabolism, local ischaemia, and necrosis. These results will have a significant negative impact on new angiogenesis [140]. On the other hand, these abnormal blood vessels have high permeability and can permeate the haemorrhagic plasma and plasma protein excessively [54, 141]. The excessive vascular permeability can easily lead to local haematoma and extravascular coagulation.

3.1. Abnormal Vascular Structure and Dysfunction

Abnormal angiogenesis, especially in tumour vessels, is characterized by network bending, hierarchical disorder, and confused blood flow; for example, some parts of the flow rate are high and other parts are stationary [142, 143], and the velocity of flow often changes [2, 144]. These malformations may be due to poor regulation of the body or incorrect angiogenesis without examination. Endothelial cells can also proliferate and migrate to other areas to form new vessels [145]. Electron microscopic studies have been reported to help determine the differences between normal and tumour blood vessels, and to report abnormal large gaps between endothelial cells and cells, as well as unusual intracellular organelles. Extravascular erythrocyte and microthrombosis are also commonly observed [146 - 148].

3.2. Abnormality of Endothelial Cell

Endothelial cells in tumour vasculature are characterised by increased exudation and abnormal intercellular junctions, which may be related to their malignant infiltration and growth [149]. Structural defects in endothelial cells may lead to blood translocation of tumour cells, which is attributed to highly active angiogenesis and microvascular remodelling [150]. In addition, endothelial cells express high levels of adhesion factors and promote cell adhesion and migration [151]. At the same time, many genes in endothelial cells of tumour tissues are abnormal.

3.3. Excessive Vascular Permeability

Vascular permeability is not only an essential physiological need to normal tissues for sustaining health, but also an important feature of many disease states, among which it is greatly increased. For example, some diseases affect angiogenesis by causing acute inflammation, such as tumours, trauma-related symptoms, and chronic inflammatory diseases [152 - 154]. Endothelial cells are more permeable in abnormal blood vessels than in normal blood vessels, and tend to exude more easily. At the same time, the intercellular connections are relaxed; these cells are usually flat [29]. Though studies on increased vascular permeability are not yet thorough, extensive vascular permeability is unlikely to be associated with normal angiogenesis; it is usually associated with pathological angiogenesis.

CONCLUDING REMARKS

Pathological angiogenesis is a symbol of wound and numbers of ischaemic, inflammatory diseases, and cancers. Angiogenesis may lead to excessive accumulation of fat in the body, and increase obesity [155]. Many inflammatory diseases have prolonged and excessive angiogenesis. Hypoxia may promote angiogenesis by stimulating the expression of related factors, thus saving myocardial infarction and prolonging the survival time of stroke patients. However, it may also lead to blindness in premature infants and diabetic patients [156]. In addition, it may also cause haemorrhagic rupture of atherosclerotic plaques and thicken the muscular layer of blood vessels, which leads to pulmonary hypertension.

In conclusion, angiogenesis was initially only associated with cancers and arthritis. However, in recent years, increasing evidence has shown that excessive, insufficient, or abnormal angiogenesis is the pathogenesis of more diseases. When angiogenesis is imbalanced, the formation of new blood vessels leads to many malignant, ischaemic, inflammatory, infectious, and immune diseases. Acquainting with the relevant pathological microenvironment will help us further understand the diseases, and conduct a vascular normalisation related research.

CONSENT FOR PUBLICATION

Not applicable.

CONFLICT OF INTEREST

The authors confirm that the contents of this chapter have no conflict of interest.

ACKNOWLEDGEMENTS

None Declare

REFERENCES

[1]   Carmeliet P. Angiogenesis in life, disease and medicine. Nature 2005; 438(7070): 932-6.
        [http://dx.doi.org/10.1038/nature04478] [PMID: 16355210]

[2]   Carmeliet P, Jain RK. Angiogenesis in cancer and other diseases. Nature 2000; 407(6801): 249-57.
[http://dx.doi.org/10.1038/35025220] [PMID: 11001068]

[3]   Polverini PJ. Angiogenesis in health and disease: insights into basic mechanisms and therapeutic opportunities. J Dent Educ 2002; 66(8): 962-75.
[http://dx.doi.org/10.1002/j.0022-0337.2002.66.8.tb03565.x] [PMID: 12214844]

[4]   Potente M, Gerhardt H, Carmeliet P. Basic and therapeutic aspects of angiogenesis. Cell 2011; 146(6): 873-87.
[http://dx.doi.org/10.1016/j.cell.2011.08.039] [PMID: 21925313]

[5]   Carmeliet P, Jain RK. Angiogenesis in cancer and other diseases. Nature 2000; 407(6801): 249.

[6]   Folkman J. Angiogenesis in cancer, vascular, rheumatoid and other disease. Nat Med 1995; 1(1): 27-31.
[http://dx.doi.org/10.1038/nm0195-27] [PMID: 7584949]

[7]   Plouët J, Shing Y. Angiogenesis. Pathol Biol (Paris) 1999; 47(4): 300.
[PMID: 10372396]

[8]   Coultas L, Chawengsaksophak K, Rossant J. Endothelial cells and VEGF in vascular development. Nature 2005; 438(7070): 937-45.
[http://dx.doi.org/10.1038/nature04479] [PMID: 16355211]

[9]   Carmeliet P, Tessier-Lavigne M. Common mechanisms of nerve and blood vessel wiring. Nature 2005; 436(7048): 193-200.
[http://dx.doi.org/10.1038/nature03875] [PMID: 16015319]

[10]   Jain RK. Molecular regulation of vessel maturation. Nat Med 2003; 9(6): 685-93.
[http://dx.doi.org/10.1038/nm0603-685] [PMID: 12778167]

[11]   Carmeliet P. Angiogenesis in health and disease. Nat Med 2003; 9(6): 653-60.
[http://dx.doi.org/10.1038/nm0603-653] [PMID: 12778163]

[12]   Alaarg A, Pérez-Medina C, Metselaar JM, *et al.* Applying nanomedicine in maladaptive inflammation and angiogenesis. Adv Drug Deliv Rev 2017; 119: 143-58.
[http://dx.doi.org/10.1016/j.addr.2017.05.009] [PMID: 28506745]

[13]   Casazza A, Laoui D, Wenes M, *et al.* Impeding macrophage entry into hypoxic tumor areas by Sema3A/Nrp1 signaling blockade inhibits angiogenesis and restores antitumor immunity. Cancer Cell 2013; 24(6): 695-709.
[http://dx.doi.org/10.1016/j.ccr.2013.11.007] [PMID: 24332039]

[14]   Christoffersson G, Vågesjö E, Vandooren J, *et al.* VEGF-A recruits a proangiogenic MMP---delivering neutrophil subset that induces angiogenesis in transplanted hypoxic tissue. Blood 2012; 120(23): 4653-62.
[http://dx.doi.org/10.1182/blood-2012-04-421040] [PMID: 22966168]

[15]   Chung AS, Wu X, Zhuang G, *et al.* An interleukin-17-mediated paracrine network promotes tumor resistance to anti-angiogenic therapy. Nat Med 2013; 19(9): 1114-23.
[http://dx.doi.org/10.1038/nm.3291] [PMID: 23913124]

[16]   Ehling J, Bartneck M, Wei X, *et al.* CCL2-dependent infiltrating macrophages promote angiogenesis in progressive liver fibrosis. Gut 2014; 63(12): 1960-71.
[http://dx.doi.org/10.1136/gutjnl-2013-306294] [PMID: 24561613]

[17]   Kolaczkowska E, Kubes P. Neutrophil recruitment and function in health and inflammation. Nat Rev Immunol 2013; 13(3): 159-75.
[http://dx.doi.org/10.1038/nri3399] [PMID: 23435331]

[18]   Stark K, Eckart A, Haidari S, *et al.* Capillary and arteriolar pericytes attract innate leukocytes exiting through venules and 'instruct' them with pattern-recognition and motility programs. Nat Immunol 2013; 14(1): 41-51.

[http://dx.doi.org/10.1038/ni.2477] [PMID: 23179077]

[19]   Silvestre JS, Mallat Z, Tedgui A, Lévy BI. Post-ischaemic neovascularization and inflammation. Cardiovasc Res 2008; 78(2): 242-9.
[http://dx.doi.org/10.1093/cvr/cvn027] [PMID: 18252762]

[20]   la Sala A, Pontecorvo L, Agresta A, Rosano G, Stabile E. Regulation of collateral blood vessel development by the innate and adaptive immune system. Trends Mol Med 2012; 18(8): 494-501.
[http://dx.doi.org/10.1016/j.molmed.2012.06.007] [PMID: 22818027]

[21]   Coussens LM, Werb Z. Inflammation and cancer. Nature 2002; 420(6917): 860-7.
[http://dx.doi.org/10.1038/nature01322] [PMID: 12490959]

[22]   Medzhitov R. Origin and physiological roles of inflammation. Nature 2008; 454(7203): 428-35.
[http://dx.doi.org/10.1038/nature07201] [PMID: 18650913]

[23]   Ross R. Atherosclerosis--an inflammatory disease. N Engl J Med 1999; 340(2): 115-26.
[http://dx.doi.org/10.1056/NEJM199901143400207] [PMID: 9887164]

[24]   Duke O, Panayi GS. The pathogenesis of rheumatoid arthritis. *Vivo.* 1988;2(1):95-103.
[PMID: 2979825]

[25]   Murdoch C, Muthana M, Lewis CE. Hypoxia regulates macrophage functions in inflammation. J Immunol 2005; 175(10): 6257-63.
[http://dx.doi.org/10.4049/jimmunol.175.10.6257] [PMID: 16272275]

[26]   Imtiyaz HZ, Williams EP, Hickey MM, *et al.* Hypoxia-inducible factor 2alpha regulates macrophage function in mouse models of acute and tumor inflammation. J Clin Invest 2010; 120(8): 2699-714.
[http://dx.doi.org/10.1172/JCI39506] [PMID: 20644254]

[27]   Strehl C, Fangradt M, Fearon U, Gaber T, Buttgereit F, Veale DJ. Hypoxia: how does the monocyte-macrophage system respond to changes in oxygen availability? J Leukoc Biol 2014; 95(2): 233-41.
[http://dx.doi.org/10.1189/jlb.1212627] [PMID: 24168857]

[28]   Bates DO, Harper SJ. Regulation of vascular permeability by vascular endothelial growth factors. Vascul Pharmacol 2002; 39(4-5): 225-37.
[http://dx.doi.org/10.1016/S1537-1891(03)00011-9] [PMID: 12747962]

[29]   Nagy JA, Benjamin L, Zeng H, Dvorak AM, Dvorak HF. Vascular permeability, vascular hyperpermeability and angiogenesis. Angiogenesis 2008; 11(2): 109-19.
[http://dx.doi.org/10.1007/s10456-008-9099-z] [PMID: 18293091]

[30]   Jackson JR, Seed MP, Kircher CH, Willoughby DA, Winkler JD. The codependence of angiogenesis and chronic inflammation. FASEB J 1997; 11(6): 457-65.
[http://dx.doi.org/10.1096/fasebj.11.6.9194526] [PMID: 9194526]

[31]   Naldini A, Carraro F. Role of inflammatory mediators in angiogenesis. Curr Drug Targets Inflamm Allergy 2005; 4(1): 3-8.
[http://dx.doi.org/10.2174/1568010053622830] [PMID: 15720228]

[32]   Lin EY, Pollard JW. Tumor-associated macrophages press the angiogenic switch in breast cancer. Cancer Res 2007; 67(11): 5064-6.
[http://dx.doi.org/10.1158/0008-5472.CAN-07-0912] [PMID: 17545580]

[33]   Owen JL, Mohamadzadeh M. Macrophages and chemokines as mediators of angiogenesis. Front Physiol 2013; 4: 159.
[http://dx.doi.org/10.3389/fphys.2013.00159] [PMID: 23847541]

[34]   Rundhaug JE. Matrix metalloproteinases and angiogenesis. J Cell Mol Med 2005; 9(2): 267-85.
[http://dx.doi.org/10.1111/j.1582-4934.2005.tb00355.x] [PMID: 15963249]

[35]   Shah PK, Falk E, Badimon JJ, *et al.* Human monocyte-derived macrophages induce collagen breakdown in fibrous caps of atherosclerotic plaques. Potential role of matrix-degrading metalloproteinases and implications for plaque rupture. Circulation 1995; 92(6): 1565-9.

[PMID: 7664441]

[36] Isner JM. Cancer and atherosclerosis: the broad mandate of angiogenesis. Circulation 1999; 99(13): 1653-5.
[http://dx.doi.org/10.1161/01.CIR.99.13.1653] [PMID: 10190871]

[37] Kwee BJ, Mooney DJ. Manipulating the intersection of angiogenesis and inflammation. Ann Biomed Eng 2015; 43(3): 628-40.
[http://dx.doi.org/10.1007/s10439-014-1145-y] [PMID: 25316589]

[38] Hunt TK, Knighton DR, Thakral KK, Goodson WH III, Andrews WS. Studies on inflammation and wound healing: angiogenesis and collagen synthesis stimulated *in vivo* by resident and activated wound macrophages. Surgery 1984; 96(1): 48-54.
[PMID: 6204395]

[39] Lin EY, Li JF, Gnatovskiy L, *et al.* Macrophages regulate the angiogenic switch in a mouse model of breast cancer. Cancer Res 2006; 66(23): 11238-46.
[http://dx.doi.org/10.1158/0008-5472.CAN-06-1278] [PMID: 17114237]

[40] Fantin A, Vieira JM, Gestri G, *et al.* Tissue macrophages act as cellular chaperones for vascular anastomosis downstream of VEGF-mediated endothelial tip cell induction. Blood 2010; 116(5): 829-40.
[http://dx.doi.org/10.1182/blood-2009-12-257832] [PMID: 20404134]

[41] Sunderkötter C, Steinbrink K, Goebeler M, Bhardwaj R, Sorg C. Macrophages and angiogenesis. J Leukoc Biol 1994; 55(3): 410-22.
[http://dx.doi.org/10.1002/jlb.55.3.410] [PMID: 7509844]

[42] Curiel TJ, Cheng P, Mottram P, *et al.* Dendritic cell subsets differentially regulate angiogenesis in human ovarian cancer. Cancer Res 2004; 64(16): 5535-8.
[http://dx.doi.org/10.1158/0008-5472.CAN-04-1272] [PMID: 15313886]

[43] Hao Q, Chen Y, Zhu Y, *et al.* Neutrophil depletion decreases VEGF-induced focal angiogenesis in the mature mouse brain. J Cereb Blood Flow Metab 2007; 27(11): 1853-60.
[http://dx.doi.org/10.1038/sj.jcbfm.9600485] [PMID: 17392691]

[44] Ohki Y, Heissig B, Sato Y, *et al.* Granulocyte colony-stimulating factor promotes neovascularization by releasing vascular endothelial growth factor from neutrophils. FASEB J 2005; 19(14): 2005-7.
[http://dx.doi.org/10.1096/fj.04-3496fje] [PMID: 16223785]

[45] Nozawa H, Chiu C, Hanahan D. Infiltrating neutrophils mediate the initial angiogenic switch in a mouse model of multistage carcinogenesis. Proc Natl Acad Sci USA 2006; 103(33): 12493-8.
[http://dx.doi.org/10.1073/pnas.0601807103] [PMID: 16891410]

[46] Norrby K. Mast cells and angiogenesis. 1985.

[47] Puxeddu I, Alian A, Piliponsky AM, Ribatti D, Panet A, Levi-Schaffer F. Human peripheral blood eosinophils induce angiogenesis. Int J Biochem Cell Biol 2005; 37(3): 628-36.
[http://dx.doi.org/10.1016/j.biocel.2004.09.001] [PMID: 15618019]

[48] Freeman MR, Schneck FX, Gagnon ML, *et al.* Peripheral blood T lymphocytes and lymphocytes infiltrating human cancers express vascular endothelial growth factor: a potential role for T cells in angiogenesis. Cancer Res 1995; 55(18): 4140-5.
[PMID: 7545086]

[49] Blotnick S, Peoples GE, Freeman MR, Eberlein TJ, Klagsbrun M. T lymphocytes synthesize and export heparin-binding epidermal growth factor-like growth factor and basic fibroblast growth factor, mitogens for vascular cells and fibroblasts: differential production and release by CD4+ and CD8+ T cells. Proc Natl Acad Sci USA 1994; 91(8): 2890-4.
[http://dx.doi.org/10.1073/pnas.91.8.2890] [PMID: 7909156]

[50] van Beem RT, Noort WA, Voermans C, *et al.* The presence of activated CD4(+) T cells is essential for the formation of colony-forming unit-endothelial cells by CD14(+) cells. J Immunol 2008; 180(7):

5141-8.
[http://dx.doi.org/10.4049/jimmunol.180.7.5141] [PMID: 18354240]

[51]   Hellingman AA, Zwaginga JJ, van Beem RT, *et al.* T-cell-pre-stimulated monocytes promote neovascularisation in a murine hind limb ischaemia model. Eur J Vasc Endovasc Surg 2011; 41(3): 418-28.
[http://dx.doi.org/10.1016/j.ejvs.2010.11.017] [PMID: 21193337]

[52]   Kasama T, Shiozawa F, Kobayashi K, *et al.* Vascular endothelial growth factor expression by activated synovial leukocytes in rheumatoid arthritis: critical involvement of the interaction with synovial fibroblasts. Arthritis Rheum 2001; 44(11): 2512-24.
[http://dx.doi.org/10.1002/1529-0131(200111)44:11<2512::AID-ART431>3.0.CO;2-O]   [PMID: 11710707]

[53]   McGovern D, Powrie F. The IL23 axis plays a key role in the pathogenesis of IBD. Gut 2007; 56(10): 1333-6.
[http://dx.doi.org/10.1136/gut.2006.115402] [PMID: 17872562]

[54]   Dvorak HF. Rous-Whipple Award Lecture. How tumors make bad blood vessels and stroma. Am J Pathol 2003; 162(6): 1747-57.
[http://dx.doi.org/10.1016/S0002-9440(10)64309-X] [PMID: 12759232]

[55]   Nagy JA, Masse EM, Herzberg KT, *et al.* Pathogenesis of ascites tumor growth: vascular permeability factor, vascular hyperpermeability, and ascites fluid accumulation. Cancer Res 1995; 55(2): 360-8.
[PMID: 7812969]

[56]   VanDeWater L, Tracy PB, Aronson D, Mann KG, Dvorak HF. Tumor cell generation of thrombin *via* functional prothrombinase assembly. Cancer Res 1985; 45(11 Pt 1): 5521-5.
[PMID: 4053025]

[57]   Bonnet CS, Walsh DA. Osteoarthritis, angiogenesis and inflammation. Rheumatology (Oxford) 2005; 44(1): 7-16.
[http://dx.doi.org/10.1093/rheumatology/keh344] [PMID: 15292527]

[58]   Stevens CR, Blake DR, Merry P, Revell PA. Levick JR. A comparative study by morphometry of the microvasculature in normal and rheumatoid synovium. Arthritis Rheumatol 2014; 34(12): 1508-13.
[http://dx.doi.org/10.1002/art.1780341206]

[59]   Haywood L, McWilliams DF, Pearson CI, *et al.* Inflammation and angiogenesis in osteoarthritis. Arthritis Rheum 2003; 48(8): 2173-7.
[http://dx.doi.org/10.1002/art.11094] [PMID: 12905470]

[60]   Giatromanolaki A, Sivridis E, Maltezos E, *et al.* Upregulated hypoxia inducible factor-1α and -2α pathway in rheumatoid arthritis and osteoarthritis. Arthritis Res Ther 2003; 5(4): R193-201.
[http://dx.doi.org/10.1186/ar756] [PMID: 12823854]

[61]   Dvorak HF, Brown LF, Detmar M, Dvorak AM. Vascular permeability factor/vascular endothelial growth factor, microvascular hyperpermeability, and angiogenesis. Am J Pathol 1995; 146(5): 1029-39.
[PMID: 7538264]

[62]   Koch AE, Burrows JC, Haines GK, Carlos TM, Harlan JM, Leibovich SJ. Immunolocalization of endothelial and leukocyte adhesion molecules in human rheumatoid and osteoarthritic synovial tissues. Lab Invest 1991; 64(3): 313-20.
[PMID: 1706003]

[63]   Fox SB, Turner GD, Gatter KC, Harris AL. The increased expression of adhesion molecules ICAM-3, E- and P-selectins on breast cancer endothelium. J Pathol 1995; 177(4): 369-76.
[http://dx.doi.org/10.1002/path.1711770407] [PMID: 8568591]

[64]   Smadja DM, Lévy BI, Silvestre J-S. Endothelial Progenitor Cells and Cardiovascular Ischemic Diseases: Characterization, Functions, and Potential Clinical Applications. Molecular Mechanisms of

Angiogenesis: From Ontogenesis to Oncogenesis. Paris: Springer Paris 2014; pp. 235-64.
[http://dx.doi.org/10.1007/978-2-8178-0466-8_11]

[65]	Silvestre JS, Smadja DM, Lévy BI. Postischemic revascularization: from cellular and molecular mechanisms to clinical applications. Physiol Rev 2013; 93(4): 1743-802.
[http://dx.doi.org/10.1152/physrev.00006.2013] [PMID: 24137021]

[66]	Dragneva G, Korpisalo P, Ylä-Herttuala S. Promoting blood vessel growth in ischemic diseases: challenges in translating preclinical potential into clinical success. Dis Model Mech 2013; 6(2): 312-22.
[http://dx.doi.org/10.1242/dmm.010413] [PMID: 23471910]

[67]	Lusis AJ. Atherosclerosis. Nature 2000; 407(6801): 233-41.
[http://dx.doi.org/10.1038/35025203] [PMID: 11001066]

[68]	Shweiki D, Neeman M, Itin A, Keshet E. Induction of vascular endothelial growth factor expression by hypoxia and by glucose deficiency in multicell spheroids: implications for tumor angiogenesis. Proc Natl Acad Sci USA 1995; 92(3): 768-72.
[http://dx.doi.org/10.1073/pnas.92.3.768] [PMID: 7531342]

[69]	Shima DT, Deutsch U, D'Amore PA. Hypoxic induction of vascular endothelial growth factor (VEGF) in human epithelial cells is mediated by increases in mRNA stability. FEBS Lett 1995; 370(3): 203-8.
[http://dx.doi.org/10.1016/0014-5793(95)00831-S] [PMID: 7656977]

[70]	Stein I, Neeman M, Shweiki D, Itin A, Keshet E. Stabilization of vascular endothelial growth factor mRNA by hypoxia and hypoglycemia and coregulation with other ischemia-induced genes. Mol Cell Biol 1995; 15(10): 5363-8.
[http://dx.doi.org/10.1128/MCB.15.10.5363] [PMID: 7565686]

[71]	Huang LE, Arany Z, Livingston DM, Bunn HF. Activation of hypoxia-inducible transcription factor depends primarily upon redox-sensitive stabilization of its alpha subunit. J Biol Chem 1996; 271(50): 32253-9.
[http://dx.doi.org/10.1074/jbc.271.50.32253] [PMID: 8943284]

[72]	Semenza GL. Transcriptional regulation by hypoxia-inducible factor 1 molecular mechanisms of oxygen homeostasis. Trends Cardiovasc Med 1996; 6(5): 151-7.
[http://dx.doi.org/10.1016/1050-1738(96)00039-4] [PMID: 21232289]

[73]	Tian H, McKnight SL, Russell DW. Endothelial PAS domain protein 1 (EPAS1), a transcription factor selectively expressed in endothelial cells. Genes Dev 1997; 11(1): 72-82.
[http://dx.doi.org/10.1101/gad.11.1.72] [PMID: 9000051]

[74]	Davis S, Aldrich TH, Jones PF, *et al.* Isolation of angiopoietin-1, a ligand for the TIE2 receptor, by secretion-trap expression cloning. Cell 1996; 87(7): 1161-9.
[http://dx.doi.org/10.1016/S0092-8674(00)81812-7] [PMID: 8980223]

[75]	Suri C, Jones PF, Patan S, *et al.* Requisite role of angiopoietin-1, a ligand for the TIE2 receptor, during embryonic angiogenesis. Cell 1996; 87(7): 1171-80.
[http://dx.doi.org/10.1016/S0092-8674(00)81813-9] [PMID: 8980224]

[76]	Vikkula M, Boon LM, Carraway KL III, *et al.* Vascular dysmorphogenesis caused by an activating mutation in the receptor tyrosine kinase TIE2. Cell 1996; 87(7): 1181-90.
[http://dx.doi.org/10.1016/S0092-8674(00)81814-0] [PMID: 8980225]

[77]	Black JE, Sirevaag AM, Greenough WT. Complex experience promotes capillary formation in young rat visual cortex. Neurosci Lett 1987; 83(3): 351-5.
[http://dx.doi.org/10.1016/0304-3940(87)90113-3] [PMID: 2450317]

[78]	Ashton N. Retinal vascularization in health and disease. Am J Ophthalmol 1957; 44(4 Pt 2): 7-17.
[http://dx.doi.org/10.1016/0002-9394(57)90426-9] [PMID: 13469948]

[79]	Dor Y, Keshet E. Ischemia-driven angiogenesis. Trends Cardiovasc Med 1997; 7(8): 289-94.
[http://dx.doi.org/10.1016/S1050-1738(97)00091-1] [PMID: 21235898]

[80]   Knighton DR, Silver IA, Hunt TK. Regulation of wound-healing angiogenesis-effect of oxygen gradients and inspired oxygen concentration. Surgery 1981; 90(2): 262-70.
[PMID: 6166996]

[81]   Pugh CW, Ratcliffe PJ. Regulation of angiogenesis by hypoxia: role of the HIF system. Nat Med 2003; 9(6): 677-84.
[http://dx.doi.org/10.1038/nm0603-677] [PMID: 12778166]

[82]   Thomlinson RH, Gray LH. The histological structure of some human lung cancers and the possible implications for radiotherapy. Br J Cancer 1955; 9(4): 539-49.
[http://dx.doi.org/10.1038/bjc.1955.55] [PMID: 13304213]

[83]   Folkman J, Merler E, Abernathy C, Williams G. Isolation of a tumor factor responsible for angiogenesis. J Exp Med 1971; 133(2): 275-88.
[http://dx.doi.org/10.1084/jem.133.2.275] [PMID: 4332371]

[84]   Knighton DR, Silver IA, Hunt TK. Regulation of wound-healing angiogenesis-effect of oxygen gradients and inspired oxygen concentration. Surgery 1981; 90(2): 262-70.
[PMID: 6166996]

[85]   Knighton DR, Hunt TK, Scheuenstuhl H, Halliday BJ, Werb Z, Banda MJ. Oxygen tension regulates the expression of angiogenesis factor by macrophages. Science 1983; 221(4617): 1283-5.
[http://dx.doi.org/10.1126/science.6612342] [PMID: 6612342]

[86]   Shweiki D, Itin A, Soffer D, Keshet E. Vascular endothelial growth factor induced by hypoxia may mediate hypoxia-initiated angiogenesis. Nature 1992; 359(6398): 843-5.
[http://dx.doi.org/10.1038/359843a0] [PMID: 1279431]

[87]   Plate KH, Breier G, Weich HA, Risau W. Vascular endothelial growth factor is a potential tumour angiogenesis factor in human gliomas *in vivo*. Nature 1992; 359(6398): 845-8.
[http://dx.doi.org/10.1038/359845a0] [PMID: 1279432]

[88]   Adair TH, Gay WJ, Montani JP. Growth regulation of the vascular system: evidence for a metabolic hypothesis. Am J Physiol 1990; 259(3 Pt 2): R393-404.
[PMID: 1697737]

[89]   Harris AL. Hypoxia--a key regulatory factor in tumour growth. Nat Rev Cancer 2002; 2(1): 38-47.
[http://dx.doi.org/10.1038/nrc704] [PMID: 11902584]

[90]   Wong BW, Marsch E, Treps L, Baes M, Carmeliet P. Endothelial cell metabolism in health and disease: impact of hypoxia. EMBO J 2017; 36(15): 2187-203.
[http://dx.doi.org/10.15252/embj.201696150] [PMID: 28637793]

[91]   Konisti S, Kiriakidis S, Paleolog EM. Hypoxia--a key regulator of angiogenesis and inflammation in rheumatoid arthritis. Nat Rev Rheumatol 2012; 8(3): 153-62.
[http://dx.doi.org/10.1038/nrrheum.2011.205] [PMID: 22293762]

[92]   Conway EM, Collen D, Carmeliet P. Molecular mechanisms of blood vessel growth. Cardiovasc Res 2001; 49(3): 507-21.
[http://dx.doi.org/10.1016/S0008-6363(00)00281-9] [PMID: 11166264]

[93]   Forsythe JA, Jiang BH, Iyer NV, *et al.* Activation of vascular endothelial growth factor gene transcription by hypoxia-inducible factor 1. Mol Cell Biol 1996; 16(9): 4604-13.
[http://dx.doi.org/10.1128/MCB.16.9.4604] [PMID: 8756616]

[94]   Gleadle JM, Ebert BL, Firth JD, Ratcliffe PJ. Regulation of angiogenic growth factor expression by hypoxia, transition metals, and chelating agents. Am J Physiol 1995; 268(6 Pt 1): C1362-8.
[http://dx.doi.org/10.1152/ajpcell.1995.268.6.C1362] [PMID: 7541940]

[95]   Goldberg MA, Schneider TJ. Similarities between the oxygen-sensing mechanisms regulating the expression of vascular endothelial growth factor and erythropoietin. J Biol Chem 1994; 269(6): 4355-9.

[PMID: 8308005]

[96]   Liu Y, Cox SR, Morita T, Kourembanas S. Hypoxia regulates vascular endothelial growth factor gene expression in endothelial cells. Identification of a 5' enhancer. Circ Res 1995; 77(3): 638-43.
       [http://dx.doi.org/10.1161/01.RES.77.3.638] [PMID: 7641334]

[97]   Melillo G, Musso T, Sica A, Taylor LS, Cox GW, Varesio L. A hypoxia-responsive element mediates a novel pathway of activation of the inducible nitric oxide synthase promoter. J Exp Med 1995; 182(6): 1683-93.
       [http://dx.doi.org/10.1084/jem.182.6.1683] [PMID: 7500013]

[98]   Ben-Yosef Y, Lahat N, Shapiro S, Bitterman H, Miller A. Regulation of endothelial matrix metalloproteinase-2 by hypoxia/reoxygenation. Circ Res 2002; 90(7): 784-91.
       [http://dx.doi.org/10.1161/01.RES.0000015588.70132.DC] [PMID: 11964371]

[99]   Kietzmann T, Roth U, Jungermann K. Induction of the plasminogen activator inhibitor-1 gene expression by mild hypoxia *via* a hypoxia response element binding the hypoxia-inducible factor-1 in rat hepatocytes. Blood 1999; 94(12): 4177-85.
       [http://dx.doi.org/10.1182/blood.V94.12.4177] [PMID: 10590062]

[100] Takahashi Y, Takahashi S, Shiga Y, Yoshimi T, Miura T. Hypoxic induction of prolyl 4-hydroxylase α (I) in cultured cells. J Biol Chem 2000; 275(19): 14139-46.
       [http://dx.doi.org/10.1074/jbc.275.19.14139] [PMID: 10799490]

[101] Meininger CJ, Schelling ME, Granger HJ. Adenosine and hypoxia stimulate proliferation and migration of endothelial cells. Am J Physiol 1988; 255(3 Pt 2): H554-62.
       [PMID: 3414822]

[102] Shreeniwas R, Ogawa S, Cozzolino F, *et al.* Macrovascular and microvascular endothelium during long-term hypoxia: alterations in cell growth, monolayer permeability, and cell surface coagulant properties. J Cell Physiol 1991; 146(1): 8-17.
       [http://dx.doi.org/10.1002/jcp.1041460103] [PMID: 1990021]

[103] Tucci M, Hammerman SI, Furfaro S, Saukonnen JJ, Conca TJ, Farber HW. Distinct effect of hypoxia on endothelial cell proliferation and cycling. Am J Physiol 1997; 272(5 Pt 1): C1700-8.
       [http://dx.doi.org/10.1152/ajpcell.1997.272.5.C1700] [PMID: 9176162]

[104] Yu F, White SB, Zhao Q, Lee FS. HIF-1alpha binding to VHL is regulated by stimulus-sensitive proline hydroxylation. Proc Natl Acad Sci USA 2001; 98(17): 9630-5.
       [http://dx.doi.org/10.1073/pnas.181341498] [PMID: 11504942]

[105] Brahimi-Horn MC, Chiche J, Pouysségur J. Hypoxia and cancer. J Mol Med (Berl) 2007; 85(12): 1301-7.
       [http://dx.doi.org/10.1007/s00109-007-0281-3] [PMID: 18026916]

[106] Rajakumar A, Doty K, Daftary A, Harger G, Conrad KP. Impaired oxygen-dependent reduction of HIF-1α and -2α proteins in pre-eclamptic placentae. Placenta 2003; 24(2-3): 199-208.
       [http://dx.doi.org/10.1053/plac.2002.0893] [PMID: 12566247]

[107] Hollander AP, Corke KP, Freemont AJ, Lewis CE. Expression of hypoxia-inducible factor 1alpha by macrophages in the rheumatoid synovium: implications for targeting of therapeutic genes to the inflamed joint. Arthritis Rheum 2001; 44(7): 1540-4.
       [http://dx.doi.org/10.1002/1529-0131(200107)44:7<1540::AID-ART277>3.0.CO;2-7]   [PMID: 11465705]

[108] Ozaki H, Yu AY, Della N, *et al.* Hypoxia inducible factor-1alpha is increased in ischemic retina: temporal and spatial correlation with VEGF expression. Invest Ophthalmol Vis Sci 1999; 40(1): 182-9.
       [PMID: 9888442]

[109] Morita M, Ohneda O, Yamashita T, *et al.* HLF/HIF-2alpha is a key factor in retinopathy of prematurity in association with erythropoietin. EMBO J 2003; 22(5): 1134-46.

[http://dx.doi.org/10.1093/emboj/cdg117] [PMID: 12606578]

[110] Majmundar AJ, Wong WJ, Simon MC. Hypoxia-inducible factors and the response to hypoxic stress. Mol Cell 2010; 40(2): 294-309.
[http://dx.doi.org/10.1016/j.molcel.2010.09.022] [PMID: 20965423]

[111] Krishnamachary B, Berg-Dixon S, Kelly B, *et al.* Regulation of colon carcinoma cell invasion by hypoxia-inducible factor 1. Cancer Res 2003; 63(5): 1138-43.
[PMID: 12615733]

[112] Iyer NV, Kotch LE, Agani F, *et al.* Cellular and developmental control of O_2 homeostasis by hypoxia-inducible factor 1 alpha. Genes Dev 1998; 12(2): 149-62.
[http://dx.doi.org/10.1101/gad.12.2.149] [PMID: 9436976]

[113] Carmeliet P, Dor Y, Herbert JM, *et al.* Role of HIF-1alpha in hypoxia-mediated apoptosis, cell proliferation and tumour angiogenesis. Nature 1998; 394(6692): 485-90.
[http://dx.doi.org/10.1038/28867] [PMID: 9697772]

[114] Ryan HE, Lo J, Johnson RS. HIF-1 alpha is required for solid tumor formation and embryonic vascularization. EMBO J 1998; 17(11): 3005-15.
[http://dx.doi.org/10.1093/emboj/17.11.3005] [PMID: 9606183]

[115] Laderoute KR, Alarcon RM, Brody MD, *et al.* Opposing effects of hypoxia on expression of the angiogenic inhibitor thrombospondin 1 and the angiogenic inducer vascular endothelial growth factor. Clin Cancer Res 2000; 6(7): 2941-50.
[PMID: 10914744]

[116] Ungvari Z, Tarantini S, Kiss T, *et al.* Endothelial dysfunction and angiogenesis impairment in the ageing vasculature. Nat Rev Cardiol 2018; 15(9): 555-65.
[http://dx.doi.org/10.1038/s41569-018-0030-z] [PMID: 29795441]

[117] Lähteenvuo J, Rosenzweig A. Effects of aging on angiogenesis. Circ Res 2012; 110(9): 1252-64.
[http://dx.doi.org/10.1161/CIRCRESAHA.111.246116] [PMID: 22539758]

[118] Ungvari Z, Podlutsky A, Sosnowska D, *et al.* Ionizing radiation promotes the acquisition of a senescence-associated secretory phenotype and impairs angiogenic capacity in cerebromicrovascular endothelial cells: role of increased DNA damage and decreased DNA repair capacity in microvascular radiosensitivity. J Gerontol A Biol Sci Med Sci 2013; 68(12): 1443-57.
[http://dx.doi.org/10.1093/gerona/glt057] [PMID: 23689827]

[119] Warrington JP, Ashpole N, Csiszar A, Lee YW, Ungvari Z, Sonntag WE. Whole brain radiation-induced vascular cognitive impairment: mechanisms and implications. J Vasc Res 2013; 50(6): 445-57.
[http://dx.doi.org/10.1159/000354227] [PMID: 24107797]

[120] Franco S, Segura I, Riese HH, Blasco MA. Decreased B16F10 melanoma growth and impaired vascularization in telomerase-deficient mice with critically short telomeres. Cancer Res 2002; 62(2): 552-9.
[PMID: 11809709]

[121] Murasawa S, Llevadot J, Silver M, Isner JM, Losordo DW, Asahara T. Constitutive human telomerase reverse transcriptase expression enhances regenerative properties of endothelial progenitor cells. Circulation 2002; 106(9): 1133-9.
[http://dx.doi.org/10.1161/01.CIR.0000027584.85865.B4] [PMID: 12196341]

[122] Asai K, Kudej RK, Shen YT, *et al.* Peripheral vascular endothelial dysfunction and apoptosis in old monkeys. Arterioscler Thromb Vasc Biol 2000; 20(6): 1493-9.
[http://dx.doi.org/10.1161/01.ATV.20.6.1493] [PMID: 10845863]

[123] Csiszar A, Ungvari Z, Koller A, Edwards JG, Kaley G. Proinflammatory phenotype of coronary arteries promotes endothelial apoptosis in aging. Physiol Genomics 2004; 17(1): 21-30.
[http://dx.doi.org/10.1152/physiolgenomics.00136.2003] [PMID: 15020720]

[124] Wagatsuma A. Effect of aging on expression of angiogenesis-related factors in mouse skeletal muscle. Exp Gerontol 2006; 41(1): 49-54.
[http://dx.doi.org/10.1016/j.exger.2005.10.003] [PMID: 16289925]

[125] Xaymardan M, Zheng J, Duignan I, *et al.* Senescent impairment in synergistic cytokine pathways that provide rapid cardioprotection in the rat heart. J Exp Med 2004; 199(6): 797-804.
[http://dx.doi.org/10.1084/jem.20031639] [PMID: 15007092]

[126] Garfinkel S, Hu X, Prudovsky IA, *et al.* FGF-1-dependent proliferative and migratory responses are impaired in senescent human umbilical vein endothelial cells and correlate with the inability to signal tyrosine phosphorylation of fibroblast growth factor receptor-1 substrates. J Cell Biol 1996; 134(3): 783-91.
[http://dx.doi.org/10.1083/jcb.134.3.783] [PMID: 8707855]

[127] Ballard VL, Edelberg JM. Harnessing hormonal signaling for cardioprotection. Sci SAGE KE 2005; 2005(51): re6.
[http://dx.doi.org/10.1126/sageke.2005.51.re6] [PMID: 16371658]

[128] Lantin-Hermoso RL, Rosenfeld CR, Yuhanna IS, German Z, Chen Z, Shaul PW. Estrogen acutely stimulates nitric oxide synthase activity in fetal pulmonary artery endothelium. Am J Physiol 1997; 273(1 Pt 1): L119-26.
[PMID: 9252548]

[129] Rubanyi GM, Johns A, Kauser K. Effect of estrogen on endothelial function and angiogenesis. Vascul Pharmacol 2002; 38(2): 89-98.
[http://dx.doi.org/10.1016/S0306-3623(02)00131-3] [PMID: 12379955]

[130] Muller FL, Lustgarten MS, Jang Y, Richardson A, Van Remmen H. Trends in oxidative aging theories. Free Radic Biol Med 2007; 43(4): 477-503.
[http://dx.doi.org/10.1016/j.freeradbiomed.2007.03.034] [PMID: 17640558]

[131] Sohal RS, Sohal BH. Hydrogen peroxide release by mitochondria increases during aging. Mech Ageing Dev 1991; 57(2): 187-202.
[http://dx.doi.org/10.1016/0047-6374(91)90034-W] [PMID: 1904965]

[132] Lee HC, Chang CM, Chi CW. Somatic mutations of mitochondrial DNA in aging and cancer progression. Ageing Res Rev 2010; 9(9) (Suppl. 1): S47-58.
[http://dx.doi.org/10.1016/j.arr.2010.08.009] [PMID: 20816876]

[133] Csiszar A, Ungvari Z, Edwards JG, *et al.* Aging-induced phenotypic changes and oxidative stress impair coronary arteriolar function. Circ Res 2002; 90(11): 1159-66.
[http://dx.doi.org/10.1161/01.RES.0000020401.61826.EA] [PMID: 12065318]

[134] Ingraham JP, Forbes ME, Riddle DR, Sonntag WE. Aging reduces hypoxia-induced microvascular growth in the rodent hippocampus. J Gerontol A Biol Sci Med Sci 2008; 63(1): 12-20.
[http://dx.doi.org/10.1093/gerona/63.1.12] [PMID: 18245756]

[135] Anversa P, Li P, Sonnenblick EH, Olivetti G. Effects of aging on quantitative structural properties of coronary vasculature and microvasculature in rats. Am J Physiol 1994; 267(3 Pt 2): H1062-73.
[PMID: 8092271]

[136] Murugesan N, Demarest TG, Madri JA, Pachter JS. Brain regional angiogenic potential at the neurovascular unit during normal aging. Neurobiol Aging. 2012;33(5):1004.e1-.e16.
[http://dx.doi.org/10.1016/j.neurobiolaging.2011.09.022]

[137] Goligorsky MS. Microvascular rarefaction: the decline and fall of blood vessels. Organogenesis 2010; 6(1): 1-10.
[http://dx.doi.org/10.4161/org.6.1.10427] [PMID: 20592859]

[138] Nagy JA, Chang S-H, Shih S-C, Dvorak AM, Dvorak HF. Heterogeneity of the tumor vasculature. Semin Thromb Hemost 2010; 36(3): 321-31.
[http://dx.doi.org/10.1055/s-0030-1253454] [PMID: 20490982]

[139]   Stubbs M, McSheehy PMJ, Griffiths JR, Bashford CL. Causes and consequences of tumour acidity and implications for treatment. Mol Med Today 2000; 6(1): 15-9.
[http://dx.doi.org/10.1016/S1357-4310(99)01615-9] [PMID: 10637570]

[140]   Semenza GL. Hypoxia-inducible factor 1: oxygen homeostasis and disease pathophysiology. Trends Mol Med 2001; 7(8): 345-50.
[http://dx.doi.org/10.1016/S1471-4914(01)02090-1] [PMID: 11516994]

[141]   Nagy JA, Feng D, Vasile E, *et al.* Permeability properties of tumor surrogate blood vessels induced by VEGF-A. Lab Invest 2006; 86(8): 767-80.
[http://dx.doi.org/10.1038/labinvest.3700436] [PMID: 16732297]

[142]   Less JR, Skalak TC, Sevick EM, Jain RK. Microvascular architecture in a mammary carcinoma: branching patterns and vessel dimensions. Cancer Res 1991; 51(1): 265-73.
[PMID: 1988088]

[143]   Patan S, Munn LL, Jain RK. Intussusceptive microvascular growth in a human colon adenocarcinoma xenograft: a novel mechanism of tumor angiogenesis. Microvasc Res 1996; 51(2): 260-72.
[http://dx.doi.org/10.1006/mvre.1996.0025] [PMID: 8778579]

[144]   Jain RK. Determinants of tumor blood flow: a review. Cancer Res 1988; 48(10): 2641-58.
[PMID: 3282647]

[145]   Munn LL. Aberrant vascular architecture in tumors and its importance in drug-based therapies. Drug Discov Today 2003; 8(9): 396-403.
[http://dx.doi.org/10.1016/S1359-6446(03)02686-2] [PMID: 12706657]

[146]   Hashizume H, Baluk P, Morikawa S, *et al.* Openings between defective endothelial cells explain tumor vessel leakiness. Am J Pathol 2000; 156(4): 1363-80.
[http://dx.doi.org/10.1016/S0002-9440(10)65006-7] [PMID: 10751361]

[147]   Dvorak AM, Kohn S, Morgan ES, Fox P, Nagy JA, Dvorak HF. The vesiculo-vacuolar organelle (VVO): a distinct endothelial cell structure that provides a transcellular pathway for macromolecular extravasation. J Leukoc Biol 1996; 59(1): 100-15.
[http://dx.doi.org/10.1002/jlb.59.1.100] [PMID: 8558058]

[148]   Warren BA, Shubik P. The growth of the blood supply to melanoma transplants in the hamster cheek pouch. Lab Invest 1966; 15(2): 464-78.
[PMID: 5932611]

[149]   Daldrup H, Shames DM, Wendland M, *et al.* Correlation of dynamic contrast-enhanced magnetic resonance imaging with histologic tumor grade: comparison of macromolecular and small-molecular contrast media. Pediatr Radiol 1998; 28(2): 67-78.
[http://dx.doi.org/10.1007/s002470050296] [PMID: 9472047]

[150]   Dvorak HF, Nagy JA, Dvorak JT, Dvorak AM. Identification and characterization of the blood vessels of solid tumors that are leaky to circulating macromolecules. Am J Pathol 1988; 133(1): 95-109.
[PMID: 2459969]

[151]   Magnussen A, Kasman IM, Norberg S, Baluk P, Murray R, McDonald DM. Rapid access of antibodies to α5β1 integrin overexpressed on the luminal surface of tumor blood vessels. Cancer Res 2005; 65(7): 2712-21.
[http://dx.doi.org/10.1158/0008-5472.CAN-04-2691] [PMID: 15805270]

[152]   Hampton T. Tumor Blood Vessels. JAMA 2007; 298(4): 394.
[http://dx.doi.org/10.1001/jama.298.4.394-c]

[153]   Dvorak HF. Rous-Whipple Award Lecture. How tumors make bad blood vessels and stroma. Am J Pathol 2003; 162(6): 1747-57.
[http://dx.doi.org/10.1016/S0002-9440(10)64309-X] [PMID: 12759232]

[154]   Nagy JA, Vasile E, Feng D, *et al.* Vascular permeability factor/vascular endothelial growth factor

induces lymphangiogenesis as well as angiogenesis. J Exp Med 2002; 196(11): 1497-506.
[http://dx.doi.org/10.1084/jem.20021244] [PMID: 12461084]

[155] Silverman KJ, Lund DP, Zetter BR, *et al.* Angiogenic activity of adipose tissue. Biochem Biophys Res Commun 1988; 153(1): 347-52.
[http://dx.doi.org/10.1016/S0006-291X(88)81229-4] [PMID: 2454107]

[156] Alon T, Hemo I, Itin A, Pe'er J, Stone J, Keshet E. Vascular endothelial growth factor acts as a survival factor for newly formed retinal vessels and has implications for retinopathy of prematurity. Nat Med 1995; 1(10): 1024-8.
[http://dx.doi.org/10.1038/nm1095-1024] [PMID: 7489357]

CHAPTER 4

Vascularization in Co-Culture Systems

Tianyi Zhang and **Xiaoxiao Cai**[*]

State Key Laboratory of Oral Diseases, West China Hospital of Stomatology, Sichuan University, Chengdu 610041, China

Abstract: Since endothelial cells are not able to create capillaries by themselves, proangiogenic factors are indispensable for endothelial cells to migrate and form microcapillaries. Thus, exogenous proangiogenic compounds are needed to improve the formation of microcapillary-like structures. Multiple forms of cell-cell interactions could result in the production of essential proangiogenic factors in co-culture systems. Many studies have examined that the co-culture systems of endothelial cells and other cell types, such as osteoblasts or mesenchymal stem cells (MSCs), can facilitate the formation of capillary-like structures. The focus of this chapter is threefold: (1) Informing the biological function of vascularization in the physiological environment. (2) Introducing typical co-culture system models for vascularization. (3) Identifying the proangiogenic factors that play crucial roles in the formation of capillary-like structures.

Keywords: Biomechanical Stimulation, Bone Tissue Engineering, Cardiac Regeneration, Cell-ECM Adhesion, Cellular Interactions, Co-culture, Cell-cell Adhesion, Direct co-culture, Endothelial Cells, Extracellular Matrix, Indirect co-culture, Media, Mesenchymal Stem Cells, Osteoblast, Oxygen Environment, Scaffolds, Seeding Methods, Skin Regeneration, Soluble Factors, Vascularization.

1. VASCULARIZATION IN PHYSIOLOGICAL AND PATHOLOGICAL CONDITIONS

1.1. Vasculogenesis and Angiogenesis in Healing and Tissue Regeneration

Vascular networks consist of two fundamental mechanisms: vasculogenesis and angiogenesis. The former is the vascular regeneration and the formation of capillary plexus *via* endothelial progenitor cells (EPCs). The latter indicates the generation of new vessels from the preexisting vascular network, involving capillary sprouting and remodeling.

[*] **Corresponding author Xiaoxiao Cai:** Sichuan University, West China School of Stomatology, China; E-mail: xcai@scu.edu.cn

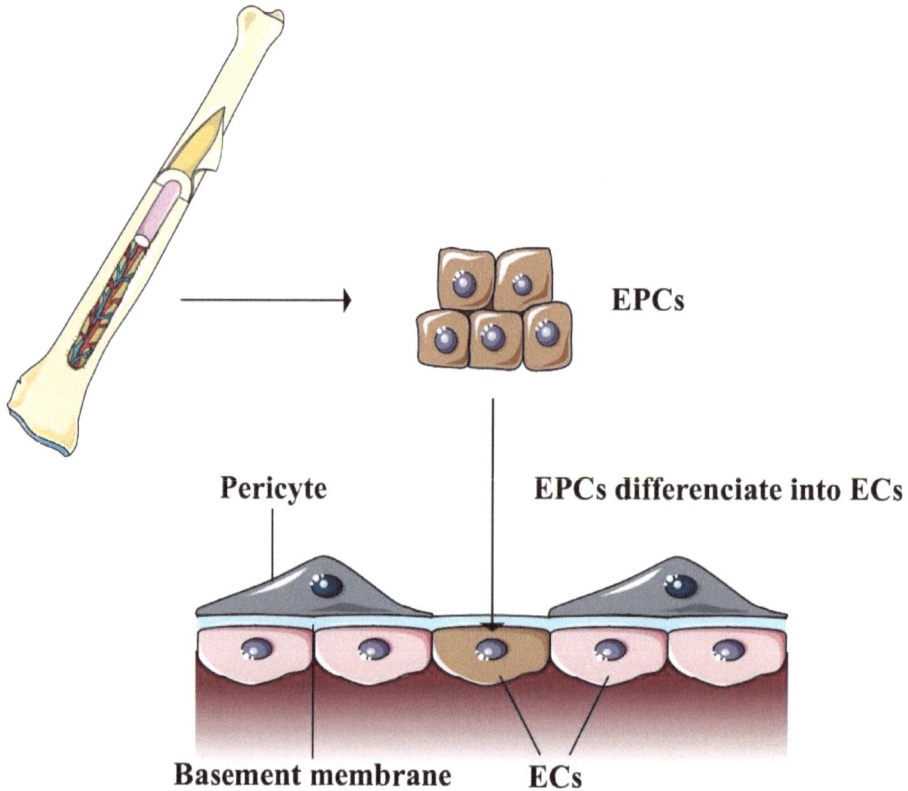

Fig. (1). Vasculogenesis.

Vascular systems play crucial roles in an abundance of biological processes, including metabolism [1], development [2], immunity [3], healing [4], and regeneration, as well as the progression of many diseases. Transporting oxygen and nutrients and removing metabolic waste, vessels are essential for growth and development. They promote the circulation of immune cells and rapidly deliver them to surrounding tissues when necessary. The versatility of vascular networks stems from the meshwork and the multiple cell types that make up blood vessels. Vessels are composed of endothelial lining surrounded by perivascular cells (PVCs) (smooth muscle cells and pericytes) and extracellular matrix (ECM) [5]. What is more, ECM is of paramount importance for normal vascular function.

Vascular networks consist of different forms of vessels: arteries, arterioles, capillaries, venules, and veins. Blood rushes from the heart into main arteries, then branches out into small arteries and flows in small arteries and capillaries. In the process of returning to the heart, blood flows through capillaries, tiny veins (venules), veins, and cavity veins. Capillaries connect small arteries and venules. The permeability of the capillary walls allows oxygen delivery and metabolic exchange. Therefore, the function of transforming nutrition and oxygen to tissues is performed mostly by microvascular networks, which encompasses small arteries, capillaries, and small veins [6].

1.1.1. Vasculogenesis

Vasculogenesis is the initial formation of vessels by cells differentiating into endothelial cells *in situ*. The process is only related to embryonic life previously, while recent studies have reported that vasculogenesis also occurs in adult tissues. The process involves angioblasts and endothelial progenitor cells.

Angioblasts, which are considered to be precursors of embryonic endothelial cells, are recruited during embryonic and fetal growth. They migrate, separate, and finally assemble into vessels, simultaneously differentiating into mature endothelial cells. Phenotypes of angioblasts change during the cardiovascular, aortic, and venous formation of embryos among multiple species [7].

EPCs usually exist in post-natal vasculogenesis, including healing, ischemia, atherosclerosis, myocardial infarction, and tumor growth. Fig. (**1**) shows the involvement of EPCs in vasculogenesis. During permanent wound healing and inflammatory reactions, such as obesity, atherosclerosis, hypercholesterolemia, and diabetes, the vasculature is distorted and confused—the phenomenon associated with a lower quantity of cEPCs(circulating endothelial progenitor cells). Vascular dysfunction manifests as a lower response to growth factors/chemokines [8]. Besides, the addition of EPCs to the injured vessel wall leads to vascular re-endothelialization. Another way to achieve re-endothelialization is colonization and re-endothelialization of cEPCs in the implanted biomaterials, which is also a hope for future therapy [9].

Neovascularization is essential to transport proteins and cells to injury sites and plays a critical role in tissue repair, growth, and development.

1.1.2. Angiogenesis

Angiogenesis refers to the formation of microvessels from existing blood vessels and capillaries through elongation, inosculation, intussusception, or sprouting.

Inosculation is vital for establishing a connection between the transplanted vascular network and the host microcirculation network. Intussusception involves bifurcation of vessels, which remodels the existing vasculature by the protrusion and fusion of opposing vessel walls in the lumen (See Fig. **2**). Compared with sprouting, intussusception angiogenesis is considerably swift in the expanded capillary network because it does not depend on cell division, but on the recombination of existing ECs [10].

Sprouting is the outcome of a multistep process that first involves the absorption of the basement membrane, followed by the infiltration and growth of ECs, the organization of lumens, and the subsequent formation of vessels and pericytes (See Fig. **3**).

Sprouting is a response to events such as inflammation and ischemia, which bring on the release of paracrine cytokines and stimulating autocrine. Subsequently, local basement membrane degradation, coupled with the migration of vascular endothelial cells guided by specialized tip cells, causes vascular leakage. Accumulation and polymerization of plasma-derived proteins, for example, fibrinogen and fibronectin, lead to the temporary ECM formation, which directs ECs to spread into the surrounding environment. The migrated endothelial cells then reassembled to form lumens and eventually mature into functional endothelial cells [11].

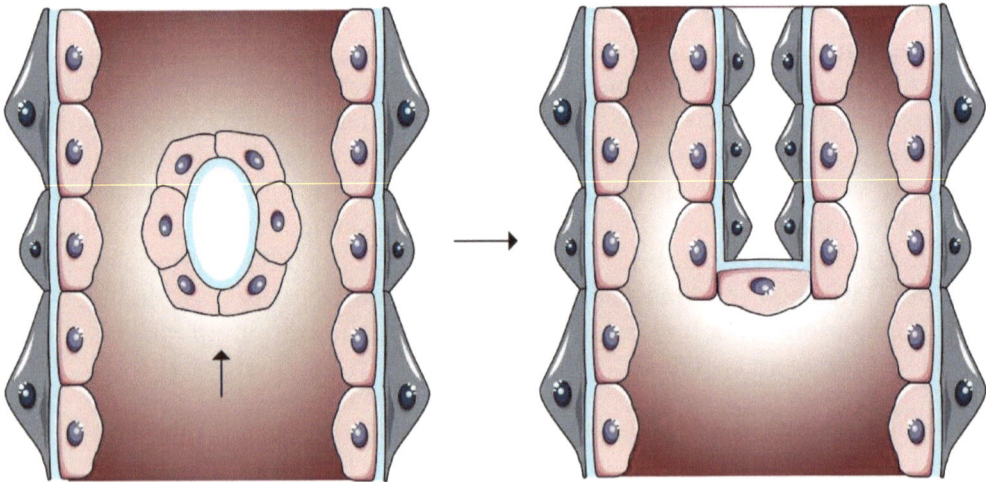

Fig. (2). Intussusception.

ECs convert into activated tip cells

Pericyte

Basement membrane   **ECs**

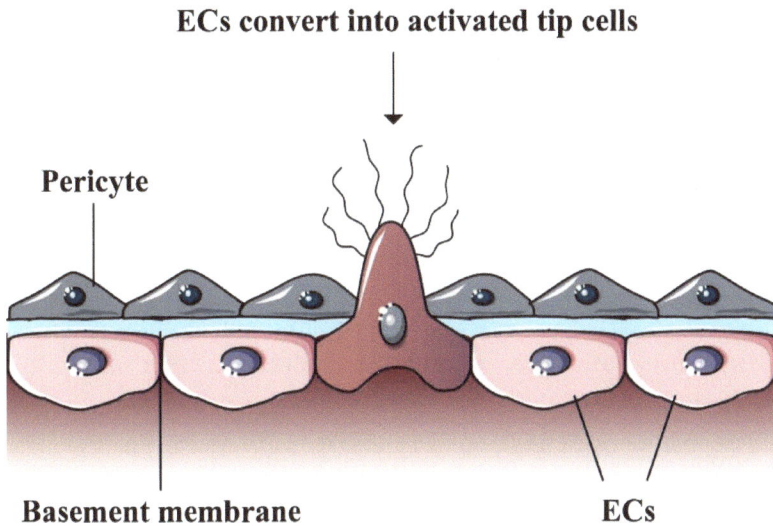

Fig. (3). Sprouting.

Stress triggers, such as injury and hypoxia, induce angiogenesis. Sprouting and intussusception form immature microvascular networks. Next, vascular networks surrounded by pericytes and cytokine-mediated cell recruitment stabilize and remodel vessel walls as well as basement membrane [12, 13]. Then, the pre-mature vascular network is trimmed and reshaped to create a sufficiently perfused, stratified vascular bed that satisfies tissue oxygenation.

In any case of healing, a local microvascular system is required to transport oxygen and nutrients to injury sites and initiate tissue regeneration. Since only layered vascular beds enable efficient blood oxygen transport and nutrient metabolism, vascular regeneration is fundamental in tissue engineering.

2. VASCULARIZATION IN PHYSIOLOGICAL MICROENVI-RONMENTS

The development of new vessels is a dynamic process, strictly regulated by angiogenesis factors and anti-angiogenic factors. The process also relies on the close interaction among endothelial cells, vascular pericytes, and ECM. Angiogenesis growth factors such as VEGF (vascular endothelial growth factor) triggers angiogenesis. Under the regulation of the Notch pathway, growth factors stimulate ECs to convert from static cells to activated tip cells. Tip cells degrade the extracellular matrix by releasing matrix metalloproteinase (MMP) and then migrate to surrounding tissues. Then proliferating endothelial stem cells form capillary buds and sprouts, which gradually grow toward angiogenic stimulus sites, and finally form lumens and combine. New blood infuses microvessels covered with pericytes and smooth muscle cells (SMCs).

2.1. Vascularization in Bone Injury Sites

When a fracture occurs, blood flows in the defect area break off, and the vascular ingrowth becomes slow. The survival rate of the central cells in broad implanted bone tissue engineering(BTE) grafts is low due to insufficient nutrient and oxygen exchange. Signaling pathways in bone regeneration are critical to achieving pleasing therapeutic efficacy [14].

In natural bone repair, endochondral ossification and intramembranous ossification occur around the vascular mesh. After the fracture occurs, ruptured blood vessels supply an anoxic environment that acts as a stimulus for upregulating several angiogenic factors.

Endochondral ossification is triggered earlier by immune reactions and inflammation, which activates a cascade of cytokine release and recruits some cell types like mesenchymal stem cells (MSCs) to the injury site.

After the formation of the vascular network, local MSCs differentiate and form cartilage substance for bone formation through resorption and mineralization.

Without the structure of cartilage, MSCs differentiate directly in the medullary cavity and initiate intramembranous ossification. Hypoxia-inducible factor 1 (HIF-1) pathway is commonly activated in the ischemic microenvironment. The activation results in the up-regulation of angiogenic factors, like VEGF [15, 16]. Afterwards, angiogenic factors trigger the migration and proliferation of endothelial cells, with tip cells serve as oxygen sensors that direct blood vessels into the hypoxic zones. In the meantime, the stalk cells proliferate and form vascular networks.

Vascular stimulating factors at bone injury sites induce osteogenesis. Studies have shown that adequate blood flow is needed to regenerate bone tissue. Endogenous VEGF is known to be essential for endochondral ossification. In rats with femoral fractures and tibial cortical bone defects, inhibition of VEGF by soluble neutralizing VEGF receptors resulted in the reduction of angiogenesis, callus mineralization and bone formation in endochondral and intramembranous ossification.

Another critical factor is the family of fibroblast growth factors. They play an integral part in the osteogenesis process. Moreover, bioactive factors also take part in signaling for the repair process. For instance, the heparan sulfate side chain can coordinate growth factors, directional regulation of cell phenotype, growth, and differentiation [17]. Heparin sulfate and its binding factors are primary biologically active elements in mineralized matrices. Deletion or deficiency of heparin sulfate leads to the destruction of the osteogenic pathway, leading to bone abnormalities.

2.2. Vascularization in Wound Healing

The main task of wound management is to restore the vascular network in the regenerative tissue for rapid healing and skin regeneration. Currently, pre-vascularized skin grafts have been developed through the cell and molecular biology [18].

The immune system plays a pivotal part in the healing process since inflammation is the initial stage of healing [19]. In the early stages of wound healing, neutrophils and macrophages hare recruited around the wound. Macrophages release signals which are inflammatory or anti-inflammatory, determining the fate of the injury [20]. During normal healing, macrophages are transformed into helping cells and release growth factors, so that the wounds enter the next stage. At this stage, granulation tissue starts to grow because of cell recruitment, especially endothelial cells and fibroblasts. In granulation tissue, new blood vessels sprout from existing blood vessels. As mentioned before, the process is called angiogenesis.

Crosstalks between endothelial cells and immune cells play a crucial part in healing, indicating their interdependence in tissue repair. Macrophages continue to provide angiogenesis and cell growth with key cytokines, and newly produced blood vessels deliver the nutrients and oxygen for cell metabolism. Chemokines secreted by macrophages can induce endothelial migration under *in vitro* conditions. Shortly, regulating immune cell behavior or adjusting immune responses may lead to breakthroughs in skin tissue engineering and wound repair. It is a significant advance in wound healing therapies that a targeted immune system to improve angiogenesis was developed recently [21].

Vascularization is a complicated procedure requiring interaction among ECM, abundant growth factors, and various cells [22]. Whereas, under certain pathophysiological conditions, such as diabetic wounds and extensive burns, normal angiogenesis is severely hampered. Artificial implants on such injuries often fail to form vascular networks, which lead to poor clinical outcomes.

Therefore, the manufacture of pre-vascularized skin grafts has become the focus of bioengineering research. Currently, co-culture is the most advanced and effective strategy for developing pre-vascularized grafts.

3. CO-CULTURE METHODOLOGIES IN TISSUE ENGINEERING

Tissue engineering is a multidisciplinary transformational technology aiming to nurture tissues to restore and enhance the function of tissues and organs damaged by trauma and disease. It uses three primary tools: scaffolds, cells, and signal molecules. The scaffold acts as a substitute for the cellular microenvironment to support tissue formation by applying biomechanical effects. Furthermore, the scaffold allows cell migration, organization, cell adhesion, and delivering binding and soluble biochemical factors. Signal molecules influence and instruct metabolism, cell phenotype, organization, and cell migration. Cells used in tissue engineering should have abundant functions, for example, synthesizing large amounts of tissue matrix, integrating with existing natural tissues, sustaining homeostasis, and assisting biochemical metabolism.

Co-culture is an experimental method to culture different cell types in a shared culture environment. It is divided into two categories: direct co-culture and indirect co-culture. In the direct co-culture, different cell types contact directly when mixed in one culture environment. In indirect co-culture, different cell types are isolated in one culture environment, in which cellular interactions occur *via* soluble factors.

3.1. Cells Types for Co-Culture Systems

3.1.1. Classified by the Role of Cells

Cell types for co-culture are classified as target cells and assisting cells. The target cells eventually form tissue and participate in tissue functions (*e.g.*, metabolism, machinery). When diverse kinds of target cells are co-cultured, every cell type act as assisting cells to help other cells. The target cells reveal an array of expected behavior with the help of assisting cells. The regulations include differentiation, proliferation, cell-cell contact, secretion of ECM to target cells, and release of signal molecules.

Co-culture systems are typically applied in tissue engineering to promote tissue formation, through multiple interactions (whether it is direct or indirect) among varied cell types. Co-culture can sustain the function of target cells in cell expansion, mainly because assisting cells usually secrete signal molecules and cytokines to regulate the medium.

The close interaction between target cells and assisting cells is often confused and cannot be achieved through exogenous control. The assisting cells are continuously monitoring and meeting the needs of target cells, efficiently providing immediate feedback to create an optimal co-culture environment.

3.1.2. Classified by Cells Types

The cell type selection is critical to managing the effects of tissue regeneration. Currently, there are three common cell types for tissue engineering: mesenchymal stem cells, endothelial cells, osteoblasts.

3.1.2.1. MSC (Mesenchymal Stem Cells)

MSCs are pluripotent cells that differentiate into various cells, such as bone, cartilage, muscle, fat, and other tissue types. MSCs are easily isolated from abundant sources such as marrow, umbilical cord blood, amniotic fluid, adipose tissue, and liver, through adherence to a plastic well surface [23]. Bone marrow is the most thoroughly studied source. Because MSCs have superiorities such as ease to separate, large-scale amplification, no differentiation, stable phenotype in culture, MSCs are very viable sources of cells in tissue engineering. Also, their low immunogenicity enables them to be applied in conditions involving potential allergens [24]. The potency of MSCs on healing has introduced promising clinical outcomes in orthopaedic diseases and defects through a variety of preclinical animal models [25 - 27].

Furthermore, the cell differentiation status should also be examined before implantation. Previous work has suggested that pre-differentiation of osteoblast-like cells in an osteoinductive medium accelerates bone formation *in vivo*, compared with undifferentiated MSCs. Some researchers compared the undifferentiated MSCs with osteoblast-loaded scaffolds in the tissue formation of rabbit calvarial defects. The results reveal that instructing MSCs to specific lineages before implantation is considerably significant, or else, the cell fate is probable to be determined directly by the host environment [28]. Nonetheless, the differentiated status of MSCs may conflict with their role as pericytes, which mostly wrap and stabilize the vessels network.

There are some other kinds of stem cells:

(1) Human umbilical cord stem cells (hUCMSCs)

hUCMSCs are collected from the human umbilical cord and are non-invasive and cheap [29]. hUCMSC is comparable to hBMSCs(human bone marrow mesenchymal stem cells) in terms of cell proliferation and differentiation [30]. Therefore, hUCMSCs are considered as robust stem cell sources for bone repair.

(2) Human umbilical cord perivascular cells (hUCPVCs)

hUCPVCs are another kind of vital MSCs detached from the human umbilical cord. Compared with hBMSCs, hUCPVCs have higher proliferation potential and can differentiate into cartilage, bone, and adipose tissue [31]. These findings encourage the potential use of hUCPVCs in bone tissue engineering.

(3) Human embryonic stem cells (hESCs)

Embryonic stem cells belong to pluripotent cell types that can differentiate into all derivatives of the three germ layers and proliferate indefinitely. However, hESCs cause clinical problems associated with tumorigenesis and immune rejection [32]. A lot of potential cell sources currently require more precise *in vitro* and *in vivo* studies to confirm the safety before clinical application.

(4) Human induced pluripotent stem cells (hiPSCs)

With the development of gene reprogramming technology, hiPSCs have become a reliable resource for stem cells in tissue regeneration [33]. Further, hiPSCs can be transducted to MSCs with enhanced proliferation, improved osteogenesis, and less tumorigenicity [34, 35].

3.1.2.2. Endothelial Cells (EC)

Angiogenesis (endothelial) cells are endothelial progenitor cells. In vascular tissue engineering, it is critical to choose the appropriate type of ECs for graft pre-vascularization. The most widely used ECs are HUVECs (human umbilical vein endothelial cells). The next commonly used ECs are HDMECs (human skin microvascular endothelial cells, usually obtained from human foreskin) [36]. Macrovascular and microvascular ECs share a lot of common phenotypes even though their structures, gene expression, and sources are distinct. These significant distinctions attribute to the type or location of endothelial cells [37 - 39].

Traditionally, mature ECs are responsible for angiogenesis after birth. Though primary endothelial cells can usually be isolated and cultured quickly, they are not the best option for tissue regeneration because of their limited cell life span. ECs are certified as phenotypic and functional heterogeneous cells. Their properties depend on the source of the endothelium, which drives more difficulties in experiments than circulating EPC populations do. Recently, endothelial-like cells (EPCs) have been extracted from blood. EPCs are promising sources of ECs for tissue engineering. Table **1** lists the differences between ECs and EPCs.

Table 1. Differences between endothelial cells and endothelial progenitor cells.

Cell Type	ECs	EPCs
Differentiative potential	Mature	Progenitor
Source	Skin Umbilical vein Adipose tissue	Peripheral blood Umbilical cord blood Bone Marrow
Location	Intima of blood vessels and lymphatic vessels	Circulating in the blood
Angiogenic efficiency	/	Higher
Proliferation speed	/	Higher

Recent researches show that peripheral blood-derived endothelial progenitor cells (PB-EPCs) proliferate ten times as much as HUVECs [40]. PB-EPCs promote fracture healing through neovascularization and increases blood vessel formation in disorders such as myocardial infarction and hindlimb ischemia [41]. Adult peripheral blood (PB) and umbilical cord blood (UCB) are the two most commonly explored origins of EPCs. The more primitive perinatal-derived UCB-EPCs have better proliferation and differentiation capabilities, compared with adult PB-EPCs. Previous investigations indicated that UCB-EPCs have at least 100 population doublings; meanwhile, maintain high levels of telomerase activity [42]. By contrast, PB-EPCs only have 20-30 populations doublings. UCB-EPCs form larger colonies earlier, undergo faster differentiation, and form capillary-like structures on Matrigel™ substrates, faster than the counterpart.

3.1.2.3. Osteoblasts

The osteoblastic lineage is mainly used in bone tissue engineering.

(1) Human osteoblasts (hOBs)

hOBs are fully mature osteoblasts detached from skeletal tissue. They are the first human autologous cells used in skeletal tissue engineering. When cultured in alpha-modified Eagle's medium (alpha-MEM) and Dulbecco's modified Eagle's medium (DMEM), hOBs may undergo osteoblast differentiation [43, 44]. Deriving the differentiation of hOBs, however, is commonly high-priced, and the cells readily differentiate after several passages.

Consequently, the stem cells with preferable differentiation potential are ideal cells for bone tissue engineering.

(2) Human bone marrow mesenchymal stem cells (hBMSCs)

hBMSCs has the potential to differentiate into various cell lines. Hence, they are recognized as the gold standard cell for tissue engineering [45]. However, obtaining hBMSCs is invasive, and the cell viability depends mainly on the patient's age and health. These deficiencies limit the application of hBMSCs.

(3) Human umbilical cord stem cells (hUCMSCs)

hUCMSCs are comparable to the gold standard hBMSCs in terms of cell proliferation, osteogenic differentiation, and synthesis of minerals.

(4) Human adipose tissue-derived mesenchymal stem cells (hAMSCs)

hAMSCs are extracted surgically from adipose tissue. Studies have shown that hAMSCs have the same potential as hBMSCs to differentiate into hOBs, synthesize and secret bone matrix [46].

(5) Human embryonic stem cells (hESCs)

As self-renewing cells, hESCs can differentiate into a range of cell lines, such as hOBs. The properties allow hESCs to become another source of stem cells in tissue engineering. When differentiating into osteogenic lineages, hESCs express osteogenic markers [47]. Similar conclusions have been reached in animal studies [48].

Additionally, fibroblasts are also widely used in co-culture systems for vascularization in tissue engineering, such as bone regeneration, skin regeneration, and skeletal muscle regeneration.

3.2. Direct Co-Culture & Indirect Co-Culture

3.2.1. Direct Co-Culture System

A direct co-culture system is a mixture of two or more different cell types. It is usually divided into two-dimensional (2D) and three-dimensional (3D) systems. A typical direct co-culture system is shown in Fig. (**4**). The 2D culture environment usually includes slides and bottles, a mixed cell population monolayer in the culture dish, and a feeder layer [49]. For instance, neonatal rat cardiomyocytes and human amniotic fluid-derived stem cells were co-cultured in a monolayer to learn their differentiation ability *in vitro* [50]. Easy and straightforward to control, 2D systems may be beneficial to studying cell behavior and interactions.

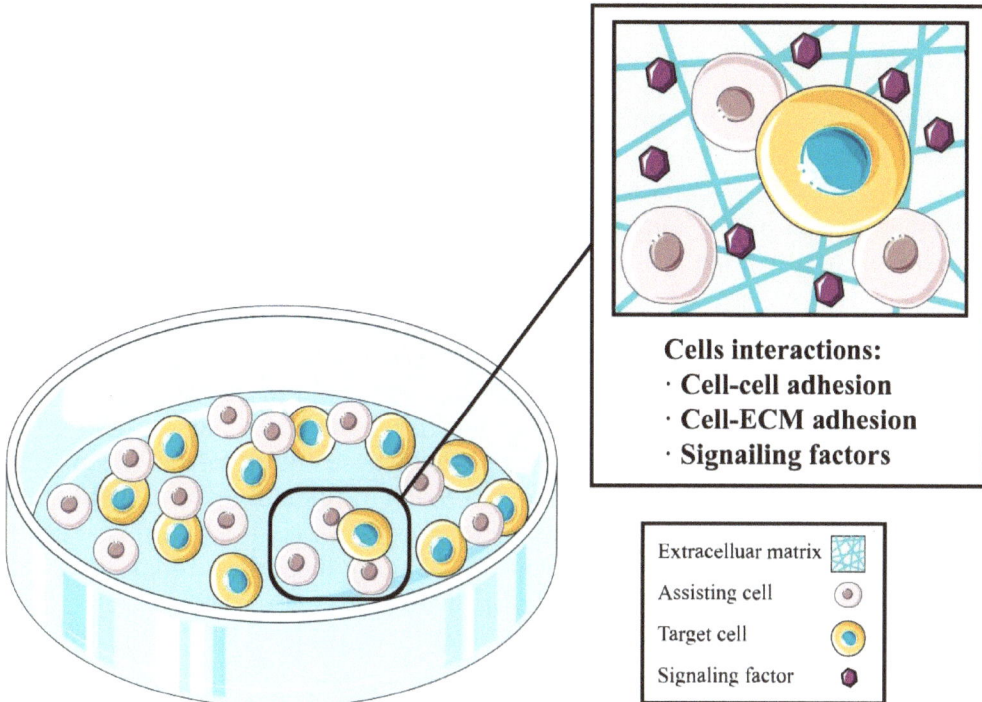

Fig. (4). A typical direct co-culture system.

By contrast, the 3D co-culture environment simulates the structure of natural tissues. That is, the mixed cell population is cultured in natural or synthetic scaffold materials, including collagen, fibrin, alginate, and agarose. That synovium-derived stem cells (SDSCs) cultivated with chondrocytes in chondrogenic medium (LG-DMEM) enhances chondrogenic potential *in vitro*, compared with the monoculture of either cell type [50].

Direct co-culture not only clarify the regeneration mechanism or promote stem cell differentiation, but also be applied in tissue repair and regeneration. In the direct co-culture systems, cell interactions involve soluble factors, paracrine signals, direct adhesion between different cells, and cell-ECM adhesion. With the presence of thrombopoietin, umbilical cord blood cells proliferate rapidly in the direct co-culture of human umbilical cord blood progenitor cells (UCB-PCs) and murine stromal cells. However, when UCB-PCs are physically detached from stromal cells, the thrombopoietin failed to promote the proliferation of them. Thus inter-cell contact plays an essential role in their behavior. Cell-cell contact and signaling are unique advantages of direct co-culture, making them available for tissue repair.

3.2.2. Indirect Co-Culture System

An indirect co-culture system generally refers to the cultivation of two or more different types and physically separated cells in the same culture environment. Two or more kinds of cells are seeded on different carriers, and then the two carriers are placed in the same culture environment so that different types of cells share the same culture system without direct contact.

Indirect co-culture systems can also be divided into 2D and 3D systems. In 2D, a transwell chamber typically achieves physical separation. A transwell chamber is a kind of permeable cup-shaped device. The permeable membrane is placed on the bottom of the cup. Generally, it is a polycarbonate film, and the pore size is 0.1 to 12.0μm. The transwell chamber is placed on the culture plate. When cell A is in the upper chamber, the components in the culture medium of cell B planted in the lower layer can affect cell A in the upper chamber (Fig. **5**). For instance, the co-culture of lipopolysaccharide-activated N9 microglia and neuronal PC12 cells was established by culturing microglia in wells with chondrocytes on a porous membrane to measure the neuroinflammatory effects [51].

In 2D co-culture, Cell B can also be cultured on a glass slide pretreated with collagen I gel and then placed in a dish seeded with cell A at a specific ratio. Besides, the co-culture of one cell culture medium(containing distinct growth factors) with another cell type is an added method of 2D indirect co-culture.

2D indirect co-cultures are manageable and straightforward, as same as monolayer cultures. Cellular communication patterns are easy to observe within a 2D indirect co-culture system. Hence, they are commonly used to study cellular interactions and behavior.

In 3D, cell types are isolated by gelatiniform encapsulation, such as hydrogel and matrigel. For instance, embryoid bodies (EBs) and cardiac fibroblasts were separately cultured in different collagen gel layers in well plates. The 3D co-culture system effectively heightened cardiomyocyte differentiation [52].

In the indirect co-culture system, paracrine signaling with soluble factors plays a primary part in the intercellular signaling, because direct contact is infeasible between physically isolated cells. A recent investigation has found that direct cell contact may not be inevitable. The percentage of beating EBs co-cultured with cardiac fibroblasts in an indirect 3D system is higher than the EBs in a monoculture [52]. This phenomenon implies that paracrine signaling is satisfactory to promote embryonic stem cells (ESCs) differentiation. Therefore, indirect co-culture can achieve tissue engineering goals while omitting direct cell-to-cell contact between different cell types, and its goals depend on the cells and medium involved in the culture.

3.3. Cellular Interactions in Co-Culture

3.3.1. Cell-Cell Adhesion

Direct cell-cell contact is usually achieved by three main categories of cell junctions: gap junctions, tight junctions, and adhesive junctions. The existence and functions of cellular connections between target cells and assisting cells have been elucidated. Gap junction channels between stem cells and terminally differentiated assisting cells allow intercellular propagation of signals, which are proved to affect stem cell differentiation. Redistribution of N-cadherin and connexin 43 that directs the gap junction formation appears in the co-culture system of human amniotic fluid-derived stem cells (hAFSCs) and neonatal cardiomyocytes. This kind of redistribution is mediated only by direct intercellular contacts in the direct co-culture, rather than indirect co-culture [50]. Therefore, the direct intercellular contacts between target cells and assisting cells provide access for intercellular signaling to achieve the desired effect.

3.3.2. Signaling via Cell-ECM Adhesion

As an environment for cell growth, ECM provides imperative stimuli for the activity and fate of target cells. The properties of ECM, such as geometry, flexibility, and the presence of mechanical signals, are required to regulate cell fate, proliferation, and migration. Changing the features of ECM usually changes cell behavior in the co-culture system.

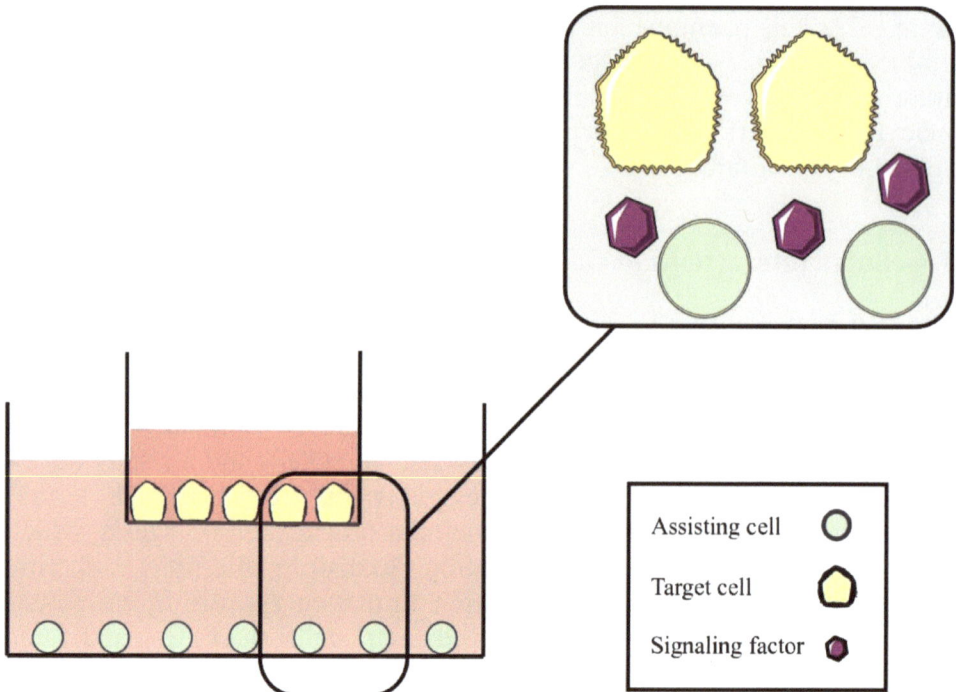

Fig. (5). Sketch map of an indirect 2D co-culture system in a transwell chamber.

Studies have shown that the perpetual remodeling of ECM alters the appearance and migration of stem cells. The assembly and degradation of ECM may have a vital influence on cell proliferation, self-renewal, and differentiation by the integrin pathway [53]. Furthermore, changes in mechanical force originated from ECM, followed by intracellular tension, which controls stem cell differentiation by balancing cytoskeletal tension and regulating RhoA-ROCK pathway [54].

There is a microenvironment or "ecological niche" around the stem cells, which supports and regulates stem cell characteristics. At the niche, several factors influence the self-renewal or differentiation of stem cells, such as the interaction inside the stem cell population, the interaction between stem cells and adjacent differentiated cells, and the interaction between stem cells and ECM [55]. Other factors, such as oxygen levels, ion concentration, the presence of growth factors and cytokines, are also relevant.

Thus, the new co-culture strategy that controls the ECM microenvironment offers new prospects for tissue engineering.

3.3.3. Paracrine Signaling Through Soluble Factors

In the co-culture system, paracrine signaling is indispensable in regulating the behavior of assisting cells and target cells. Mediated through soluble factors, paracrine mainly acts as a meaningful role in the indirect co-culture systems. When co-cultured with fibroblasts, MSCs were induced to differentiate toward VSMC(vascular smooth muscle cell)-like cells with the enhanced functional activity of VSMC phenotype, increased transcription of marker genes, upregulated expression of contractile apparatus proteins [56]. Moreover, in an indirect 3D co-culture system, the ligament fibroblasts can promote MSCs to differentiate into fibroblasts [57]. The source of assisting cells and the physical-chemical properties of the matrix influence the differentiation of target cells [58].

When stem cells act as assisting cells, the soluble factors they excrete regulates the behavior of target cells. In an indirect co-culture system, bone marrow-derived MSCs regulate the migration, gene expression, and proliferation of skin fibroblasts [59]. Since soluble factors have critical effects on secreting ECM proteins and growth factors, Adipose-derived stem cells (ADSCs) enhance the healing ability of human dermal fibroblasts [60]. Furthermore, MSCs exhibit immunomodulatory effects by cytokines that inhibit T cell secretion, and also secrete prostaglandin E directly [61].

In a co-culture system, remote cell signaling enables assisting cells in supporting and regulating target cells. Additionally, cellular interactions can be beneficial to both parties. For example, a co-culture system involving human synovium-derived stem cells and chondrocytes drives the differentiation of MSCs toward the chondrocyte lineage and reduces MSC hypertrophy [62]. Another study informs that the co-culture of meniscal cells and MSCs not only improve the production of meniscus ECM but also reduced MSC hypertrophy [63]. The value of two-way cellular interactions in tissue engineering is noteworthy, and both cell populations benefit from co-culture systems.

3.4. Factors in Co-Culture Systems

3.4.1. Culture Media

The medium should be carefully selected when maintaining cell viability in cell type and tissue growth. Optimal conditions should induce the formation of vascular network and bone and avert unnecessary inducing differentiation of other cells. The heterogeneity of the experimental parameters in the study makes it tough to contrast different co-culture systems. In a study of multiple media types (with/without growth supplements and with/without osteogenic inducers) aimed at

MSC/EPC co-culture, separately applied endothelial supplements and bone media enhance angiogenesis and mineralization. Previous work has suggested that bone media have a higher ability to induce MSC to differentiate into osteoblasts, compared with other media types [64]. When MSCs were incubated in EGM2(Endothelial Cell Growth Medium-2), microvessel formation efficiency increased, compared with bone media supplemented without growth factor. Due to the low survival rate of endothelial cells, few researchers explored co-culture with osteoinductive media. However, this result conflicts with another experiment. Liu *et al.* designed a vascular network structure within 3D osteogenic grafts *in vitro*, indicating that even without angiogenic supplements, the osteoinductive mediator itself completely supports the survival and distribution of ECs in a short period [65]. They also showed that despite the lack of osteogenic regulation in EGM, osteoinductive mediators could efficiently induce osteogenic differentiation of MSCs. Growth factors added to the endothelial growth medium promote endothelial differentiation efficiently. Prior studies have approved that the selection of medium and soluble mediators, such as osteogenic inducers and angiogenic growth factors, has much influence over cell survival and differentiation.

3.4.2. Seeding Methods

Although the best method has not been determined, the 3D scaffold structure has several different cell inoculation methods. Some factors need to be considered, for example, how to culture the cells before implantation, whether the cells should be cultured in the optimal medium or separately, whether to use a dynamic culture system, the incubation time, and which inoculation ratio to use to allow effective cell interactions and crosstalk.

Table **2** outlines the probable advantages and disadvantages of several seeding methods.

3.4.3. Extracellular Matrix (ECM)

When applied to some specific matrix, cells can be retained in the porous medium in the scaffold for efficient seeding. ECM and scaffolds both play crucial parts in localizing growth factors, migration, adhesion, and cell survival. Nevertheless, we usually use some appropriate matrix to block the interaction between different cells. Some researchers used matrigel to immobilize co-cultures [66]. Since matrigel contains growth factors that mask the effects of co-culture, they cannot reveal the property and interaction effects of co-cultures. Besides, matrigel is obtained from the basement membrane matrix produced by mouse tumor cell lines, so it is not feasible in a clinical setting. In contrast, the application of fibrin has been established and commonly applied in a variety of clinical application scenarios. Fibrin has ideal physical properties for biodegradability, biocompatibility, and wound healing because it comes from the ingredients of normal wound repair. Possibly it will replace ECM matrices in future co-culture systems.

Table 2. Advantages and disadvantages of different seeding methods of co-culture in vascularized tissue engineering.

Seeding Method	Advantages	Disadvantages
Mixing distinct cells directly in co-culture container	Easy to observe Easy to use Evenly distributed mixed cells	Confirm the survival of each type of cells in the same medium before combining the co-culture system
Seeding on the same scaffold, first by one cell type; second by another type	Better local environment to support cell growth and differentiation	Asymmetric distribution of cells Vascularization depending on the design of scaffolds Less intercellular communication
Seeding different cell types in different scaffolds separately, then combining	Preserving cellular phenotype and characteristics	
Co-culture followed by biochemical stimulation	Advancing cell viability and survival rate Evenly distributed mixed cells	Challenging experimental design

Researchers seek to develop a wider variety of co-culture models in order further to intensify control of cell-cell interactions and cell distribution. Techniques for producing co-cultures models comprise soft lithography, patterning based on dielectric electrophoretic, patterning based on the switchable surface in 2D layered systems [67]. The application of 3D patterned co-culture may create a more evenly distributed vascular network in the osteogenic implant, thereby maximizing the effects of heterotypic interactions and soluble factors.

3.4.4. Oxygen Environment

The fracture destroys blood supply, which generates the up-regulation of various angiogenic genes and hypoxia in the injury sites. Experimentally, HIF-1a and the activated pathways assist bone development through stimulating bone progenitor cell differentiation, promoting bone formation and remodeling, and combining angiogenesis mechanisms [68]. More robust bone comes into being when VEGF is upregulated in osteoblasts, followed by continuous activation of HIF-1a. In contrast, inhibition of HIF-1a results in less vascularized bones. In order to repair CSD(critical size defect) in rat skulls, HIF-1a is overexpressed with lentivirus in bone marrow-derived MSCs, resulting in improved vascularization and bone repair compared with the control group.

A prior investigation discussed the effects of hypoxic cultures on MSC/EPC based on the principle that anoxia is responsible for angiogenesis. However, compared with normoxic culture, angiogenesis is blocked after hypoxia stimulation, which indicates the necessity of pre-vascularized structures. Further research is required to strengthen the significance of hypoxic stimulation for a co-culture system to promote tissue regeneration and tissue repair responses *in vivo*.

3.4.5. 3D Scaffolds

A three-dimensional scaffold is applied to provide cells with a local environment. Moreover, as a temporary matrix, it can sustain cellular differentiation and growth and provide mechanical inductions for growth factor expression. The optimal scaffold is supposed to have mechanical stability to support the injury sites temporarily and reabsorb at a proper speed that matches the formation of new bone in the injury sites.

Cells can identify the structural characteristics of the environment and then differentiate respectively in tissue remodeling. Because the scaffold is designed as a mimic of the microenvironment of healthy tissues *in vivo*, the structural features of the scaffold are essential in guiding cell distribution, adhesion, and differentiation. The trabecular structure locates in the spongy central region, and an extension of the vascular network supports the bone in the outer cortical region. The porous network and trabecular microstructure are both required to encourage vascular growth and tissue formation. Most importantly, the host tissue interacts with the biological material and develop the required cellular response to help tissue repair.

The type of material, pore size, mechanical strength, manufacturability, surface chemical properties, and space properties should be considered as primary factors in scaffold designing.

3.4.6. Biomechanical Stimulation

Bioreactors facilitate the homogenous distribution of MSCs and efficient mass transport in the scaffold. Additionally, the signs at the beginning of some pathways appear in normal physiological conditions (*e.g.*, homeostasis, tissue formation, and tissue development) and positively affect osteogenesis in the graft [54, 69].

Several lines of evidence suggest that the dynamic culture of MSC-based tissue-engineered grafts increase cell viability in grafts and promote mineralization [70, 71]. The selection of different dynamic systems influences seeding, proliferation, and differentiation of cells [72]. In another investigation, researchers measured mRNA expression levels to evaluate the optimal strain amplitude for the differentiation of MSCs into tenocytes. The results indicate that appropriate mechanical stimulations can induce the differentiation of MSCs towards tenocytes [73].

Multiple studies have proved the success of bone defect repair by co-culturing the osteoprogenitor-like EPCs and MSCs on scaffolds without mechanical stimulation [74]. However, the results only meet short-term and medium-term goals in bone formation and may, therefore, limit late bone formation and long-term remodeling. One prior research reported a dynamic co-culture system in cell-encapsulated alginate microspheres [75]. In this study, researchers compared the co-culture of human osteoblasts and HUVECs with the single culture. Regretfully, the effect of dynamic culture on the co-culture system for vascularized tissue engineering has not been studied before.

Owing to the mechanically sensitive properties of MSCs, osteoblasts, endothelial cells, and related physiology, biomechanical stimulation is imperative to promote the differentiation of target cells and assisting cells in co-culture systems.

4. APPLICATIONS OF CO-CULTURE IN VASCULAR ENGINEERING

Because the versatility of blood vessels is closely related to multiple cell types that make up blood vessels, co-culture systems are usually used in vascularization projects for different purposes. Except for endothelial cells, other cell types directly or indirectly related to the circulatory system can be applied to vascularize in different biological materials. Compared with single-culture EC systems, co-culture is closer to *in vivo* conditions and helps to form more complex vasculature. This section aims to portray the experiments of co-culture systems for vascular engineering under *in vivo* conditions.

4.1. Bone Tissue Engineering (BTE)

Bone is a blood-rich organ that relies heavily on blood vessels to maintain cell survival and bone integrity. The bone tissue gets sufficient nutrients and oxygen from blood vessels within the proximity of 100-200μm. As a result, vascular insufficiency often occurs when repairing defects using other tissue engineering grafts or thicker tissues.

For bone regeneration, the early formation of the blood vessel is a fundamental process. Capillary-like structures *in vitro* can be quickly integrated into normal functioning blood vessels after implantation *in vivo*. Endothelial cells have the potential to produce capillary-like structures in some particular environment, which could promote vascularization of bones. However, vascular structures created by endothelial cells are often weak and labile [76]. Prior research found that the specific pro-angiogenic factors could not be created by endothelial cells to promote the vascular maturation [77], while endothelial-osteoblastic direct contact and communication is a prerequisite for capillaries and bone formation. Therefore, the superiority of the co-culture of osteoblasts and endothelial cells are revealed in 2D or 3D models.

To confirm osteogenesis and angiogenesis through animal co-culture, Liu *et al.* implanted hiPSC-MSC and hUVEC together in a Calcium phosphate cement (CPC) scaffold [78]. Four groups of artificial skull defects were tested in nude rats: (1) CPC alone (CPC control); (2) hUVEC-seeded CPC (CPC-hUVEC); (3) hiPSC-MSC-seeded CPC (CPC-hiPSC-MSC); (4) hUVEC co-cultured with hiPSC-MSCs. Three months later, the co-culture system formed a wider area of new bone in all groups, with a percentage of 46.38% ± 3.8% (p <0.05), which is more than four times the percentage of CPC control that is 10.61% ± 1.43. As a result, the synergistic effect existed between hUVEC and hiPSC-MSC has been confirmed in the field of bone regeneration.

Another experiment co-cultured hUVECs and diverse mesenchymal stem cells in CPC scaffolds to assess angiogenesis and osteogenesis [79]. Nude rats with skull defects were divided in six groups: (1) CPC (Control 1); (2) CPC with hBMSCs (Control 2); (3) hUCMSCs + hUVECs; (4) hiPSC-MSCs + hUVECs; (5) hESC-MSCs + hUVECs; (6) hBMSCs + hUVECs. The findings indicate that the blood vessel density and bone density of co-culture group were higher than those of the control group (p <0.05). Thus, in addition to hBMSCs, the hiPSC-MSCs, hESC-MSCs and hUCMSCs could be co-cultured with hUVECs in the CPC scaffold, which also displayed the remarkable angiogenic and osteogenic potential *in vivo*.

4.2. Cardiac Regeneration

The heart is one of the organs with the highest rate of blood vessels to (myocardial) cells. Therefore, vascularization is very important for highly metabolic tissue, such as heart tissue. Because adult cardiomyocytes are terminally differentiated, their proliferation rate is inferior, and cardiac regeneration is minimal.

4.2.1. Use of Endothelial Cells and MSCs

Since there are few sources of human microvascular wall cells, MSCs are commonly used to enhance vascularization in tissue-engineered implants. They can be derived from various tissues and expanded relatively easily. Since they are available from clinical patients, they also have translational potential. When co-cultured with ECs, MSCs can act as microvascular endothelial cells and stabilize the structure of ECs through paracrine when co-cultured with ECs.

In a related study, researchers co-cultured HUVECs and mesenchymal precursor cells (MPCs) in the ratio of 4:1 in three-dimensional fibronectin-collagen gel and then put the construct into cranial windows of SCID (immunodeficient) mice. In the first 14 days after implantation of the structure, the number of perfused vessels in blood circulation increased rapidly, and then remained stable and functioning normally for four months after implantation. The results were contrary to constructs inoculated with HUVECs alone. Although the early morphological changes of the two constructs were similar, HUVECs alone had the minimal perfusion and finally vanished after two months. In fibronectin-containing collagen scaffolds, it takes four days for hMSCs co-cultured with HUVECs in a 4:1 ratio to generate a stable vascular network. Then on day 11, the implant was perfused sufficiently by host vasculature after transplanted subcutaneously in immunodeficient mice [80].

In another study, a tri-culture system on a micro-mode fibrin scaffold was designed by researchers, which concluded neonatal rat cardiomyocytes, rat aortic endothelial cells or vascular endothelial cells, and neonatal rat fibroblasts in a ratio of 4:2:1 [81]. The scaffold is a 27μm microporous network structure, surrounded by the evenly distributed 60m microchannels; its mechanical stiffness is approximate to natural heart tissue (70-90kPa). The speed of degradation of scaffolds dropped with the application of factor XIII (FXIII) and/or protease inhibitor (aprotinin), compared with the rapid degradation of unmodified scaffolds, both *in vivo* and *in vitro*. Then a non-ischemic adult rat was selected by researchers to be implanted the tri-culture scaffold into its ventricular wall, which reduced collagen deposition after seven days. Implants that were unmodified or only applied with aprotinin could not be located after two weeks, while two of eight implants with FXIII were discovered, but severely degraded. The construct supports the survival and growth of cells *in vitro* and promotes the formation of EC-lined lumen-like tubular structures in the construct channels as well as co-localization of viable cardiomyocytes and ECs.

An experiment using neonatal rat cardiomyocytes in the simulated 3D structure showed a connection between cell viability and oxygen gradient. In particular, cardiomyocytes are highly sensitive to hypoxia [82]. Pre-vascularization of the engineered heart tissue increases blood perfusion after implantation, ensuring cell survival and physiological function of engineered heart tissue. A nude mouse model study involves scaffold-free cardiac patches composed of fibroblasts, ECs, as well as cardiomyocytes, which improved myocardial cell viability, whereas a single-culture of cardiomyocytes did not survive. That HUVECs, mouse embryonic fibroblasts, and hESC-CMs co-cultured on poly-1-lactic acid/polyglycolic acid scaffolds, also increased myocardial cell survival rate [83].

4.2.2. Use of Human Pluripotent Stem Cells (HPSCs)

Besides the typical microvascular structure *in vivo* was consist of different sources of PVCs as well as ECs in a naturally occurring matrix, a bicellular vascular population in an artificial matrix could also be obtained by applying human pluripotent stem cells. Early-stage vascular cells can be induced from these hPSCs to convert to ECs or pericytes and form a tubular network of HA hyaluronic acid within three days *in vitro*. After two weeks, subcutaneous implantation of these two-cell populations of HA hydrogels in nude mice results in neovascularization, integration with host blood circulation, and blood perfusion.

4.2.3. Use of Endothelial Cells and Perivascular Cells

PVCs support the ECs and maintain the stability of vasculature [84]. Therefore, the inclusion of PVCs in an engineering organization may improve its durability and functionality. Various models reveal the interaction between ECs and PVCs, which promote angiogenesis in engineered tissues.

Vascular smooth muscle cells (SMCs) are also used for angiogenesis in engineered tissues. Immunodeficient mice are implanted subcutaneously in the abdominal wall with collagen fibronectin gel carrying Bcl-2-HUVEC to form chimeric vessels containing human ECs and host VSMCs at eight weeks [85]. The Bcl-2-HUVEC and human aortic SMC(2:1) co-transplanted in a polyglycolic acid scaffold could be averted to SMC-mediated gel shrinkage to promote the maturation of vessels and enhance the vitality of engineered tissue [86].

4.3. Skin Regeneration

Under pathophysiological conditions, such as diabetic wounds or extensive burns, the natural angiogenesis process is severely impeded. Artificial implants on such wounds often form a vascular network unsuccessfully and have poor clinical results. Making pre-vascularized skin grafts is preferable to solve the problem. With the aid of vascular networks, bioengineered tissues have been well-developed. Pre-vascularization strategies, for instance, increasing growth factor delivery, using bioactive materials, have been discovered. Up to now, co-culture systems are the most effective strategy for developing pre-vascularized skin grafts.

An earlier study suggested that the co-culture of HUVECs and human diploid fibroblasts (containing 2% fetal bovine serum) in human adult dermis resulted in vascular formation. Microvascular-like structures appeared on day seven; complex capillary networks appeared on day 14. The number of vascular structures analogous to microvascular beds was furtherly increased by the joining of VEGF, while the anti-angiogenic drug suramin and anti-VEGF antibodies both inhibited angiogenesis [87].

Researchers also vascularize tissue-engineered constructs with endothelial colony-forming cells through a mixture of growth factors (recombinant human VEGF165, SDF-1, TNF-α, angiopoietin-1 and bFGF) grafts to accelerate the healing of diabetic wounds [88]. Among the growth factors mentioned above, VEGF secreted by macrophages is a critical vascular growth factor and a chemokine in wound repair.

Endothelial cells forming capillary walls undergo crosstalk with fibroblasts. The co-culture of ECs and fibroblasts was proved to be favorable in the vascularized dermo-epidermal skin grafts. When co-cultured human dermal fibroblasts (HDFs) and human skin microvascular endothelial cells (HDMECs) in three-dimensional hydrogels, the vascular network was formed within three weeks [18]. HDMECs cannot form a capillary structure when it was cultured alone in the hydrogel, demonstrating the importance of the co-culture system. Further experiments found that skin grafts with lymphatic vessels and vascular networks that were transplanted into the wounds of nude mice can connect with existing blood vessels and form a functional skin layer. The preformed blood vessels could also be applied to excrete the wound exudate of bioengineered grafts, thereby accelerating wound repair.

By co-culturing HUVECs, human skin fibroblasts, and keratinocytes (1: 1: 1) for 31 days in a collagen scaffold, skin tissue was vascularized. The endothelialized skin equivalents built capillary-like structures *in vitro*. The capillary-like structures formation is ascribed to the cooperation between ECs, fibroblasts, and the complex ECM synthesized by fibroblasts [89]. Besides, some other cell types also regulated the capillary-like structures. For instance, human epithelial cells can stabilize and regulate capillary formation in the endothelialized tissue-engineered human skin [90].

The co-culture of HUVECs and fibroblasts could be applied in vascularization of other tissue, too. For instance, a 3D multiplex culture system comprising ECs (HUVECs or hESCECs), mouse myoblasts on porous cells, and mouse embryonic fibroblasts can induce the formation of endothelial vascular networks in engineered skeletal muscle tissue.

CONSENT FOR PUBLICATION

Not applicable.

CONFLICT OF INTEREST

The authors confirm that the contents of this chapter have no conflict of interest.

ACKNOWLEDGEMENTS

This study was supported by National Key R&D Program of China (2019YFA0110600) and National Natural Science Foundation of China (81970986, 81771125).

REFERENCES

[1]   Potente M, Gerhardt H, Carmeliet P. Basic and therapeutic aspects of angiogenesis. Cell 2011; 146(6): 873-87.
[http://dx.doi.org/10.1016/j.cell.2011.08.039] [PMID: 21925313]

[2]   Fish JE, Wythe JD. The molecular regulation of arteriovenous specification and maintenance. Dev Dyn 2015; 244(3): 391-409.
[http://dx.doi.org/10.1002/dvdy.24252] [PMID: 25641373]

[3]   Bauer SM, Bauer RJ, Velazquez OC. Angiogenesis, vasculogenesis, and induction of healing in chronic wounds. Vasc Endovascular Surg 2005; 39(4): 293-306.
[http://dx.doi.org/10.1177/153857440503900401] [PMID: 16079938]

[4]   Park KM, Gerecht S. Harnessing developmental processes for vascular engineering and regeneration. Development 2014; 141(14): 2760-9.
[http://dx.doi.org/10.1242/dev.102194] [PMID: 25005471]

[5]   Wanjare M, Kusuma S, Gerecht S. Perivascular cells in blood vessel regeneration. Biotechnol J 2013; 8(4): 434-47.
[http://dx.doi.org/10.1002/biot.201200199] [PMID: 23554249]

[6]   Staiculescu MC, Foote C, Meininger GA, Martinez-Lemus LA. The role of reactive oxygen species in microvascular remodeling. Int J Mol Sci 2014; 15(12): 23792-835.
[http://dx.doi.org/10.3390/ijms151223792] [PMID: 25535075]

[7]   Ema M, Rossant J. Cell fate decisions in early blood vessel formation. Trends Cardiovasc Med 2003; 13(6): 254-9.
[http://dx.doi.org/10.1016/S1050-1738(03)00105-1] [PMID: 12922023]

[8]   Velazquez OC. Angiogenesis and vasculogenesis: inducing the growth of new blood vessels and wound healing by stimulation of bone marrow-derived progenitor cell mobilization and homing. J Vasc Surg 2007; 45(Suppl A): A39-47.

[9]   Hirschi KK, Ingram DA, Yoder MC. Assessing identity, phenotype, and fate of endothelial progenitor cells. Arterioscler Thromb Vasc Biol 2008; 28(9): 1584-95.
[http://dx.doi.org/10.1161/ATVBAHA.107.155960] [PMID: 18669889]

[10]   Burri PH, Hlushchuk R, Djonov V. Intussusceptive angiogenesis: its emergence, its characteristics, and its significance. Dev Dyn 2004; 231(3): 474-88.
[http://dx.doi.org/10.1002/dvdy.20184] [PMID: 15376313]

[11]   Mercado-Pagán AE, Stahl AM, Shanjani Y, Yang Y. Vascularization in bone tissue engineering constructs. Ann Biomed Eng 2015; 43(3): 718-29.
[http://dx.doi.org/10.1007/s10439-015-1253-3] [PMID: 25616591]

[12]   Kant RJ, Coulombe KLK. Integrated approaches to spatiotemporally directing angiogenesis in host and engineered tissues. Acta Biomater 2018; 69: 42-62.
[http://dx.doi.org/10.1016/j.actbio.2018.01.017] [PMID: 29371132]

[13]   Cochain C, Channon KM, Silvestre JS. Angiogenesis in the infarcted myocardium. Antioxid Redox Signal 2013; 18(9): 1100-13.
[http://dx.doi.org/10.1089/ars.2012.4849] [PMID: 22870932]

[14]   Liu Y, Chan JK, Teoh SH. Review of vascularised bone tissue-engineering strategies with a focus on co-culture systems. J Tissue Eng Regen Med 2015; 9(2): 85-105.
[http://dx.doi.org/10.1002/term.1617] [PMID: 23166000]

[15]   Coulon C, Georgiadou M, Roncal C, De Bock K, Langenberg T, Carmeliet P. From vessel sprouting to normalization: role of the prolyl hydroxylase domain protein/hypoxia-inducible factor oxygen-sensing machinery. Arterioscler Thromb Vasc Biol 2010; 30(12): 2331-6.
[http://dx.doi.org/10.1161/ATVBAHA.110.214106] [PMID: 20966400]

[16]   Germain S, Monnot C, Muller L, Eichmann A. Hypoxia-driven angiogenesis: role of tip cells and extracellular matrix scaffolding. Curr Opin Hematol 2010; 17(3): 245-51.
[http://dx.doi.org/10.1097/MOH.0b013e32833865b9] [PMID: 20308893]

[17]   Cool SM, Nurcombe V. Heparan sulfate regulation of progenitor cell fate. J Cell Biochem 2006; 99(4): 1040-51.
[http://dx.doi.org/10.1002/jcb.20936] [PMID: 16767693]

[18]   Marino D, Luginbühl J, Scola S, Meuli M, Reichmann E. Bioengineering dermo-epidermal skin grafts with blood and lymphatic capillaries. Sci Transl Med 2014; 6(221): 221ra14.
[http://dx.doi.org/10.1126/scitranslmed.3006894] [PMID: 24477001]

[19]   Eaton KV, Yang HL, Giachelli CM, Scatena M. Engineering macrophages to control the inflammatory response and angiogenesis. Exp Cell Res 2015; 339(2): 300-9.
[http://dx.doi.org/10.1016/j.yexcr.2015.11.021] [PMID: 26610863]

[20]   Kloc M, Ghobrial RM, Wosik J, Lewicka A, Lewicki S, Kubiak JZ. Macrophage functions in wound healing. J Tissue Eng Regen Med 2019; 13(1): 99-109.
[PMID: 30445662]

[21]   Castellano D, Sanchis A, Blanes M, *et al.* Electrospun poly(hydroxybutyrate) scaffolds promote engraftment of human skin equivalents*via*macrophage M2 polarization and angiogenesis. J Tissue Eng Regen Med 2018; 12(2): e983-94.
[http://dx.doi.org/10.1002/term.2420] [PMID: 28111928]

[22]   Lisovsky A, Chamberlain MD, Wells LA, Sefton MV. Cell interactions with vascular regenerative MAA-based materials in the context of wound healing. Adv Healthc Mater 2015; 4(16): 2375-87.
[http://dx.doi.org/10.1002/adhm.201500192] [PMID: 26010569]

[23]　Pittenger MF, Mackay AM, Beck SC, *et al.* Multilineage potential of adult human mesenchymal stem cells. Science 1999; 284(5411): 143-7.
[http://dx.doi.org/10.1126/science.284.5411.143] [PMID: 10102814]

[24]　Zhang ZY, Teoh SH, Hui JH, Fisk NM, Choolani M, Chan JK. The potential of human fetal mesenchymal stem cells for off-the-shelf bone tissue engineering application. Biomaterials 2012; 33(9): 2656-72.
[http://dx.doi.org/10.1016/j.biomaterials.2011.12.025] [PMID: 22217806]

[25]　Jia Y, Qiu S, Xu J, Kang Q, Chai Y. Exosomes secreted by young mesenchymal stem cells promote new bone formation during distraction osteogenesis in older rats. Calcif Tissue Int 2020; 106(5): 509-17.
[http://dx.doi.org/10.1007/s00223-019-00656-4] [PMID: 32103287]

[26]　Meinel L, Betz O, Fajardo R, *et al.* Silk based biomaterials to heal critical sized femur defects. Bone 2006; 39(4): 922-31.
[http://dx.doi.org/10.1016/j.bone.2006.04.019] [PMID: 16757219]

[27]　Wang P, Zhang Z. Bone marrow-derived mesenchymal stem cells promote healing of rabbit tibial fractures*via*JAK-STAT signaling pathway. Exp Ther Med 2020; 19(3): 2310-6.
[http://dx.doi.org/10.3892/etm.2020.8441] [PMID: 32104299]

[28]　Schantz JT, Woodruff MA, Lam CX, *et al.* Differentiation potential of mesenchymal progenitor cells following transplantation into calvarial defects. J Mech Behav Biomed Mater 2012; 11: 132-42.
[http://dx.doi.org/10.1016/j.jmbbm.2012.02.008] [PMID: 22658162]

[29]　Zhao L, Weir MD, Xu HH. An injectable calcium phosphate-alginate hydrogel-umbilical cord mesenchymal stem cell paste for bone tissue engineering. Biomaterials 2010; 31(25): 6502-10.
[http://dx.doi.org/10.1016/j.biomaterials.2010.05.017] [PMID: 20570346]

[30]　Chen W, Liu J, Manuchehrabadi N, Weir MD, Zhu Z, Xu HH. Umbilical cord and bone marrow mesenchymal stem cell seeding on macroporous calcium phosphate for bone regeneration in rat cranial defects. Biomaterials 2013; 34(38): 9917-25.
[http://dx.doi.org/10.1016/j.biomaterials.2013.09.002] [PMID: 24054499]

[31]　Kajiyama S, Ujiie Y, Nishikawa S, *et al.* Bone formation by human umbilical cord perivascular cells. J Biomed Mater Res A 2015; 103(8): 2807-14.
[http://dx.doi.org/10.1002/jbm.a.35396] [PMID: 25676366]

[32]　Zhao T, Zhang ZN, Rong Z, Xu Y. Immunogenicity of induced pluripotent stem cells. Nature 2011; 474(7350): 212-5.
[http://dx.doi.org/10.1038/nature10135] [PMID: 21572395]

[33]　Nakagawa M, Koyanagi M, Tanabe K, *et al.* Generation of induced pluripotent stem cells without Myc from mouse and human fibroblasts. Nat Biotechnol 2008; 26(1): 101-6.
[http://dx.doi.org/10.1038/nbt1374] [PMID: 18059259]

[34]　Wang P, Liu X, Zhao L, *et al.* Bone tissue engineering*via*human induced pluripotent, umbilical cord and bone marrow mesenchymal stem cells in rat cranium. Acta Biomater 2015; 18: 236-48.
[http://dx.doi.org/10.1016/j.actbio.2015.02.011] [PMID: 25712391]

[35]　Zhao Q, Gregory CA, Lee RH, *et al.* MSCs derived from iPSCs with a modified protocol are tumor-tropic but have much less potential to promote tumors than bone marrow MSCs. Proc Natl Acad Sci USA 2015; 112(2): 530-5.
[http://dx.doi.org/10.1073/pnas.1423008112] [PMID: 25548183]

[36]　Peters K, Schmidt H, Unger RE, Otto M, Kamp G, Kirkpatrick CJ. Software-supported image quantification of angiogenesis in an *in vitro* culture system: application to studies of biocompatibility. Biomaterials 2002; 23(16): 3413-9.
[http://dx.doi.org/10.1016/S0142-9612(02)00042-X] [PMID: 12099284]

[37]　Langenkamp E, Molema G. Microvascular endothelial cell heterogeneity: general concepts and

pharmacological consequences for anti-angiogenic therapy of cancer. Cell Tissue Res 2009; 335(1): 205-22.
[http://dx.doi.org/10.1007/s00441-008-0642-4] [PMID: 18677515]

[38] Minami T, Muramatsu M, Kume T. Organ/tissue-specific vascular endothelial cell heterogeneity in health and disease. Biol Pharm Bull 2019; 42(10): 1609-19.
[http://dx.doi.org/10.1248/bpb.b19-00531] [PMID: 31582649]

[39] Nolan DJ, Ginsberg M, Israely E, *et al.* Molecular signatures of tissue-specific microvascular endothelial cell heterogeneity in organ maintenance and regeneration. Dev Cell 2013; 26(2): 204-19.
[http://dx.doi.org/10.1016/j.devcel.2013.06.017] [PMID: 23871589]

[40] Murohara T. Cord blood-derived early outgrowth endothelial progenitor cells. Microvasc Res 2010; 79(3): 174-7.
[http://dx.doi.org/10.1016/j.mvr.2010.01.008] [PMID: 20085776]

[41] Dimmeler S, Zeiher AM. Cell therapy of acute myocardial infarction: open questions. Cardiology 2009; 113(3): 155-60.
[http://dx.doi.org/10.1159/000187652] [PMID: 19122455]

[42] Ingram DA, Mead LE, Tanaka H, *et al.* Identification of a novel hierarchy of endothelial progenitor cells using human peripheral and umbilical cord blood. Blood 2004; 104(9): 2752-60.
[http://dx.doi.org/10.1182/blood-2004-04-1396] [PMID: 15226175]

[43] Coelho MJ, Cabral AT, Fernande MH. Human bone cell cultures in biocompatibility testing. Part I: osteoblastic differentiation of serially passaged human bone marrow cells cultured in alpha-MEM and in DMEM. Biomaterials 2000; 21(11): 1087-94.
[http://dx.doi.org/10.1016/S0142-9612(99)00284-7] [PMID: 10817260]

[44] Coelho MJ, Fernandes MH. Human bone cell cultures in biocompatibility testing. Part II: effect of ascorbic acid, beta-glycerophosphate and dexamethasone on osteoblastic differentiation. Biomaterials 2000; 21(11): 1095-102.
[http://dx.doi.org/10.1016/S0142-9612(99)00192-1] [PMID: 10817261]

[45] Baksh D, Yao R, Tuan RS. Comparison of proliferative and multilineage differentiation potential of human mesenchymal stem cells derived from umbilical cord and bone marrow. Stem Cells 2007; 25(6): 1384-92.
[http://dx.doi.org/10.1634/stemcells.2006-0709] [PMID: 17332507]

[46] Marini F, Luzi E, Fabbri S, *et al.* Osteogenic differentiation of adipose tissue-derived mesenchymal stem cells on nanostructured Ti6Al4V and Ti13Nb13Zr. Clin Cases Miner Bone Metab 2015; 12(3): 224-37.
[http://dx.doi.org/10.11138/ccmbm/2015.12.3.224] [PMID: 26811701]

[47] Chen W, Zhou H, Weir MD, Tang M, Bao C, Xu HH. Human embryonic stem cell-derived mesenchymal stem cell seeding on calcium phosphate cement-chitosan-RGD scaffold for bone repair. Tissue Eng Part A 2013; 19(7-8): 915-27.
[http://dx.doi.org/10.1089/ten.tea.2012.0172] [PMID: 23092172]

[48] Liu X, Wang P, Chen W, Weir MD, Bao C, Xu HH. Human embryonic stem cells and macroporous calcium phosphate construct for bone regeneration in cranial defects in rats. Acta Biomater 2014; 10(10): 4484-93.
[http://dx.doi.org/10.1016/j.actbio.2014.06.027] [PMID: 24972090]

[49] Nishiofuku M, Yoshikawa M, Ouji Y, *et al.* Modulated differentiation of embryonic stem cells into hepatocyte-like cells by coculture with hepatic stellate cells. J Biosci Bioeng 2011; 111(1): 71-7.
[http://dx.doi.org/10.1016/j.jbiosc.2010.08.005] [PMID: 20801713]

[50] Guan X, Delo DM, Atala A, Soker S. *in vitro* cardiomyogenic potential of human amniotic fluid stem cells. J Tissue Eng Regen Med 2011; 5(3): 220-8.
[http://dx.doi.org/10.1002/term.308] [PMID: 20687122]

[51]   Renaud J, Martinoli MG. Development of an insert co-culture system of two cellular types in the absence of cell-cell contact. J Vis Exp 2016; (113):

[52]   Ou DB, He Y, Chen R, *et al.* Three-dimensional co-culture facilitates the differentiation of embryonic stem cells into mature cardiomyocytes. J Cell Biochem 2011; 112(12): 3555-62.
[http://dx.doi.org/10.1002/jcb.23283] [PMID: 21780160]

[53]   Daley WP, Peters SB, Larsen M. Extracellular matrix dynamics in development and regenerative medicine. J Cell Sci 2008; 121(Pt 3): 255-64.
[http://dx.doi.org/10.1242/jcs.006064] [PMID: 18216330]

[54]   Dado D, Sagi M, Levenberg S, Zemel A. Mechanical control of stem cell differentiation. Regen Med 2012; 7(1): 101-16.
[http://dx.doi.org/10.2217/rme.11.99] [PMID: 22168501]

[55]   Singh A, Yadav CB, Tabassum N, Bajpeyee AK, Verma V. Stem cell niche: Dynamic neighbor of stem cells. Eur J Cell Biol 2019; 98(2-4): 65-73.
[http://dx.doi.org/10.1016/j.ejcb.2018.12.001] [PMID: 30563738]

[56]   Li N, Sanyour H, Remund T, Kelly P, Hong Z. Vascular extracellular matrix and fibroblasts-coculture directed differentiation of human mesenchymal stem cells toward smooth muscle-like cells for vascular tissue engineering. Mater Sci Eng C 2018; 93: 61-9.
[http://dx.doi.org/10.1016/j.msec.2018.07.061] [PMID: 30274093]

[57]   Fan H, Liu H, Toh SL, Goh JC. Enhanced differentiation of mesenchymal stem cells co-cultured with ligament fibroblasts on gelatin/silk fibroin hybrid scaffold. Biomaterials 2008; 29(8): 1017-27.
[http://dx.doi.org/10.1016/j.biomaterials.2007.10.048] [PMID: 18023476]

[58]   Philp D, Chen SS, Fitzgerald W, Orenstein J, Margolis L, Kleinman HK. Complex extracellular matrices promote tissue-specific stem cell differentiation. Stem Cells 2005; 23(2): 288-96.
[http://dx.doi.org/10.1634/stemcells.2002-0109] [PMID: 15671151]

[59]   Smith AN, Willis E, Chan VT, *et al.* Mesenchymal stem cells induce dermal fibroblast responses to injury. Exp Cell Res 2010; 316(1): 48-54.
[http://dx.doi.org/10.1016/j.yexcr.2009.08.001] [PMID: 19666021]

[60]   Kim WS, Park BS, Sung JH, *et al.* Wound healing effect of adipose-derived stem cells: a critical role of secretory factors on human dermal fibroblasts. J Dermatol Sci 2007; 48(1): 15-24.
[http://dx.doi.org/10.1016/j.jdermsci.2007.05.018] [PMID: 17643966]

[61]   Aggarwal S, Pittenger MF. Human mesenchymal stem cells modulate allogeneic immune cell responses. Blood 2005; 105(4): 1815-22.
[http://dx.doi.org/10.1182/blood-2004-04-1559] [PMID: 15494428]

[62]   Bian L, Zhai DY, Mauck RL, Burdick JA. Coculture of human mesenchymal stem cells and articular chondrocytes reduces hypertrophy and enhances functional properties of engineered cartilage. Tissue Eng Part A 2011; 17(7-8): 1137-45.
[http://dx.doi.org/10.1089/ten.tea.2010.0531] [PMID: 21142648]

[63]   Cui X, Hasegawa A, Lotz M, D'Lima D. Structured three-dimensional co-culture of mesenchymal stem cells with meniscus cells promotes meniscal phenotype without hypertrophy. Biotechnol Bioeng 2012; 109(9): 2369-80.
[http://dx.doi.org/10.1002/bit.24495] [PMID: 22422555]

[64]   Dohle E, Fuchs S, Kolbe M, Hofmann A, Schmidt H, Kirkpatrick CJ. Sonic hedgehog promotes angiogenesis and osteogenesis in a coculture system consisting of primary osteoblasts and outgrowth endothelial cells. Tissue Eng Part A 2010; 16(4): 1235-7.
[http://dx.doi.org/10.1089/ten.tea.2009.0493] [PMID: 19886747]

[65]   Liu Y, Teoh SH, Chong MS, *et al.* Vasculogenic and osteogenesis-enhancing potential of human umbilical cord blood endothelial colony-forming cells. Stem Cells 2012; 30(9): 1911-24.
[http://dx.doi.org/10.1002/stem.1164] [PMID: 22761003]

[66]   Koob S, Torio-Padron N, Stark GB, Hannig C, Stankovic Z, Finkenzeller G. Bone formation and neovascularization mediated by mesenchymal stem cells and endothelial cells in critical-sized calvarial defects. Tissue Eng Part A 2011; 17(3-4): 311-21.
[http://dx.doi.org/10.1089/ten.tea.2010.0338] [PMID: 20799886]

[67]   Kaji H, Camci-Unal G, Langer R, Khademhosseini A. Engineering systems for the generation of patterned co-cultures for controlling cell-cell interactions. Biochim Biophys Acta 2011; 1810(3): 239-50.
[http://dx.doi.org/10.1016/j.bbagen.2010.07.002] [PMID: 20655984]

[68]   Hulley PA, Bishop T, Vernet A, *et al.* Hypoxia-inducible factor 1-alpha does not regulate osteoclastogenesis but enhances bone resorption activity*via*prolyl-4-hydroxylase 2. J Pathol 2017; 242(3): 322-33.
[http://dx.doi.org/10.1002/path.4906] [PMID: 28418093]

[69]   McCoy RJ, O'Brien FJ. Influence of shear stress in perfusion bioreactor cultures for the development of three-dimensional bone tissue constructs: a review. Tissue Eng Part B Rev 2010; 16(6): 587-601.
[http://dx.doi.org/10.1089/ten.teb.2010.0370] [PMID: 20799909]

[70]   Carvalho MS, Silva JC, Udangawa RN, *et al.* Co-culture cell-derived extracellular matrix loaded electrospun microfibrous scaffolds for bone tissue engineering. Mater Sci Eng C 2019; 99: 479-90.
[http://dx.doi.org/10.1016/j.msec.2019.01.127] [PMID: 30889723]

[71]   Zhang ZY, Teoh SH, Chong WS, *et al.* A biaxial rotating bioreactor for the culture of fetal mesenchymal stem cells for bone tissue engineering. Biomaterials 2009; 30(14): 2694-704.
[http://dx.doi.org/10.1016/j.biomaterials.2009.01.028] [PMID: 19223070]

[72]   Zhang ZY, Teoh SH, Teo EY, *et al.* A comparison of bioreactors for culture of fetal mesenchymal stem cells for bone tissue engineering. Biomaterials 2010; 31(33): 8684-95.
[http://dx.doi.org/10.1016/j.biomaterials.2010.07.097] [PMID: 20739062]

[73]   Morita Y, Sato T, Higashiura K, *et al.* The optimal mechanical condition in stem cell-to-tenocyte differentiation determined with the homogeneous strain distributions and the cellular orientation control. Biol Open 2019; 8(5): bio039164.
[http://dx.doi.org/10.1242/bio.039164] [PMID: 31118166]

[74]   Seebach C, Henrich D, Kähling C, *et al.* Endothelial progenitor cells and mesenchymal stem cells seeded onto beta-TCP granules enhance early vascularization and bone healing in a critical-sized bone defect in rats. Tissue Eng Part A 2010; 16(6): 1961-70.
[http://dx.doi.org/10.1089/ten.tea.2009.0715] [PMID: 20088701]

[75]   Grellier M, Granja PL, Fricain JC, *et al.* The effect of the co-immobilization of human osteoprogenitors and endothelial cells within alginate microspheres on mineralization in a bone defect. Biomaterials 2009; 30(19): 3271-8.
[http://dx.doi.org/10.1016/j.biomaterials.2009.02.033] [PMID: 19299013]

[76]   Koike N, Fukumura D, Gralla O, Au P, Schechner JS, Jain RK. Tissue engineering: creation of long-lasting blood vessels. Nature 2004; 428(6979): 138-9.
[http://dx.doi.org/10.1038/428138a] [PMID: 15014486]

[77]   Unger RE, Dohle E, Kirkpatrick CJ. Improving vascularization of engineered bone through the generation of pro-angiogenic effects in co-culture systems. Adv Drug Deliv Rev 2015; 94: 116-25.
[http://dx.doi.org/10.1016/j.addr.2015.03.012] [PMID: 25817732]

[78]   Liu X, Chen W, Zhang C, *et al.* Co-seeding human endothelial cells with human-induced pluripotent stem cell-derived mesenchymal stem cells on calcium phosphate scaffold enhances osteogenesis and vascularization in rats. Tissue Eng Part A 2017; 23(11-12): 546-55.
[http://dx.doi.org/10.1089/ten.tea.2016.0485] [PMID: 28287922]

[79]   Chen W, Liu X, Chen Q, *et al.* Angiogenic and osteogenic regeneration in rats *via* calcium phosphate scaffold and endothelial cell co-culture with human bone marrow mesenchymal stem cells (MSCs),

human umbilical cord MSCs, human induced pluripotent stem cell-derived MSCs and human embryonic stem cell-derived MSCs. J Tissue Eng Regen Med 2018; 12(1): 191-203.
[http://dx.doi.org/10.1002/term.2395] [PMID: 28098961]

[80]   Tsigkou O, Pomerantseva I, Spencer JA, *et al.* Engineered vascularized bone grafts. Proc Natl Acad Sci USA 2010; 107(8): 3311-6.
[http://dx.doi.org/10.1073/pnas.0905445107] [PMID: 20133604]

[81]   Thomson KS, Korte FS, Giachelli CM, Ratner BD, Regnier M, Scatena M. Prevascularized microtemplated fibrin scaffolds for cardiac tissue engineering applications. Tissue Eng Part A 2013; 19(7-8): 967-77.
[http://dx.doi.org/10.1089/ten.tea.2012.0286] [PMID: 23317311]

[82]   Huang S, Yang Y, Yang Q, Zhao Q, Ye X. Engineered circulatory scaffolds for building cardiac tissue. J Thorac Dis 2018; 10 (Suppl. 20): S2312-28.
[http://dx.doi.org/10.21037/jtd.2017.12.92] [PMID: 30123572]

[83]   Lesman A, Habib M, Caspi O, *et al.* Transplantation of a tissue-engineered human vascularized cardiac muscle. Tissue Eng Part A 2010; 16(1): 115-25.
[http://dx.doi.org/10.1089/ten.tea.2009.0130] [PMID: 19642856]

[84]   Geevarghese A, Herman IM. Pericyte-endothelial crosstalk: implications and opportunities for advanced cellular therapies. Transl Res 2014; 163(4): 296-306.
[http://dx.doi.org/10.1016/j.trsl.2014.01.011] [PMID: 24530608]

[85]   Enis DR, Shepherd BR, Wang Y, *et al.* Induction, differentiation, and remodeling of blood vessels after transplantation of Bcl-2-transduced endothelial cells. Proc Natl Acad Sci USA 2005; 102(2): 425-30.
[http://dx.doi.org/10.1073/pnas.0408357102] [PMID: 15625106]

[86]   Shepherd BR, Jay SM, Saltzman WM, Tellides G, Pober JS. Human aortic smooth muscle cells promote arteriole formation by coengrafted endothelial cells. Tissue Eng Part A 2009; 15(1): 165-73.
[http://dx.doi.org/10.1089/ten.tea.2008.0010] [PMID: 18620481]

[87]   Bishop ET, Bell GT, Bloor S, Broom IJ, Hendry NF, Wheatley DN. An *in vitro* model of angiogenesis: basic features. Angiogenesis 1999; 3(4): 335-44.
[http://dx.doi.org/10.1023/A:1026546219962] [PMID: 14517413]

[88]   Shen YI, Cho H, Papa AE, *et al.* Engineered human vascularized constructs accelerate diabetic wound healing. Biomaterials 2016; 102: 107-19.
[http://dx.doi.org/10.1016/j.biomaterials.2016.06.009] [PMID: 27328431]

[89]   Hudon V, Berthod F, Black AF, Damour O, Germain L, Auger FA. A tissue-engineered endothelialized dermis to study the modulation of angiogenic and angiostatic molecules on capillary-like tube formation *in vitro*. Br J Dermatol 2003; 148(6): 1094-104.
[http://dx.doi.org/10.1046/j.1365-2133.2003.05298.x] [PMID: 12828735]

[90]   Rochon MH, Fradette J, Fortin V, *et al.* Normal human epithelial cells regulate the size and morphology of tissue-engineered capillaries. Tissue Eng Part A 2010; 16(5): 1457-68.
[http://dx.doi.org/10.1089/ten.tea.2009.0090] [PMID: 19938961]

CHAPTER 5

Angiogenesis and Materials

Xin Qin, Yuting Wen and **Xiaoxiao Cai**[*]

State Key Laboratory of Oral Diseases, National Clinical Research Center for Oral Diseases, West China Hospital of Stomatology, Sichuan University, China

Abstract: In the process of tissue engineering vascularization, as one of the three major factors of tissue engineering, scaffold materials play a vital role in the various processes of vascularization. As a starting point, this chapter first introduces the reader to several strategies for scaffolding materials to promote vascularization. It basically describes the reasons for the use of scaffold materials in vascularization and the necessity of scaffold materials in tissue engineering vascularization. Then we will focus on how to properly use the scaffold material. The design of the scaffold material itself is a key factor in the function of the material, and the scaffold material with ideal biological properties can make the process of vascularization more effective. Factors such as the topology of the material and the physical and chemical properties of the material affect the success rate of vascularization to varying degrees. We hope that readers can obtain the basic knowledge and principles of stent design from this chapter. Finally, a number of fresh ideas have emerged for the design of tissue engineered vascular materials, such as new material handling methods, new ways of combining cells, and so on, which have improved the vascularization process to varying degrees. Scaffold materials have shown attractive prospects and great possibilities in vascular tissue engineering. Previous studies have found many materials associated with vascularization, but there are also many problems to be solved. With the development of materials science and engineering, it is believed that there will be new vascular stent materials with better performance and more suitable for vascularization in the future.

Keywords: Angiogenesis, Degradation Modes of Scaffold, Materials, Tissue Engineering Vascularization.

1. INTRODUCTION

With the development of tissue engineering technology, repairing large-area bone defects using tissue-engineered bone has become a widely used approach. The scaffold materials with good three-dimensional structures can promote cell growth and proliferation and tissue ingrowth, and a variety of materials can reach a combined effect to meet the clinical demand.

[*] **Corresponding author Xiaoxiao Cai:** Sichuan University, West China School of Stomatology, China; E-mail: xcai@scu.edu.cn

Xiaoxiao Cai (Ed.)

In addition, it is important to actively seek new material preparation technology and improve the existing methods, in order to create an excellent scaffold. However, vascularization is still a major challenge for bone tissue engineering.

2. TISSUE ENGINEERING BIOLOGICAL SCAFFOLD MATERIALS

The ideal healthy blood vessels in tissue engineering should have the following conditions: They should possess the ability to or simulate *in vivo* conditions. There are three layers of vascular wall structure, namely adventitia, media, and intima. They should exhibit biocompatibility, that is to say, it is not easy to produce thrombus and immune rejection. At the same time, it has similar biological characteristics with normal blood vessels, such as relaxation response to drug stimulation and needs to possess the mechanics exhibited by blood vessels. Third, it should also exhibit viscoelasticity and can withstand certain pressure. Scaffold materials are indispensable for achieving good vascularization processes.

With the continuous development of tissue engineering, repairing large-area bone defects using tissue-engineered bone has become a widely used approach. The scaffold materials with good three-dimensional structures can promote cell growth and proliferation, tissue growth, osteogenesis, and vascularization. Each scaffold has its own inadequacies; therefore, the combination of a variety of materials can achieve a combined effect to meet the clinical demand. In addition, it is important to actively seek new material preparation technology and improve the existing methods, in order to create excellent scaffolds [1 - 4].

2.1. The Design Concept of Ideal Biological Scaffold Material

Vascular stent materials play an indispensable role in vascular tissue engineering. The desired support for the growth of cells *in vitro*, which can provide the organization of blood vessels, certain mechanical strength and mechanical properties, seed cell growth. The ideal stent should have the following properties: (1) Controllable rate of biodegradation; (2) low immunogenicity, which does not cause an inflammatory response; (3) good biological properties; (4) good mechanical and physiological properties; (5) suitable porous structure; and (6) they should be easy to process and sterilize [1]. In short, it is necessary to provide specific three-dimensional (tubular) scaffolds for the construction of tissue-engineered blood vessels, which can be used as matrix materials for the implantation of vascular cells, so that the inoculated cells can be positioned, attached, and localized growth and proliferation can be promoted. Meanwhile, the materials can arrange the cells in the space of scaffolds, differentiate with specific functions, and synthesize appropriate extracellular matrix, following which tissue-engineered blood vessels can be transplanted. *in vivo* stent materials should also have strong learning support and anti-blood pressure functions [2, 5 - 7].

2.2. Classification and Basic Concept of Biological Scaffold Materials

Currently used vascular stent materials include natural biological stent materials, synthetic biodegradable polymer materials, composite materials, and nanomaterials (Table **1**).

Table 1. Comparison of various scaffolds materials.

Material Classification	Natural Scaffold Materials	Synthetic Biomaterials	Nanomaterials
Material composition	Macromolecular materials such as collagen, alginate, *etc.*; acellular tissue matrix materials such as acellular dermal matrix, acellular vascular matrix, acellular bladder matrix, *etc.*	Non degradable polymer materials: Polyester and expanded polytetrafluoroethylene; degradable polymer materials: Poly (β - hydroxybutyric acid), poly (aminic acid), polycarbonate, polyurethane, poly (lactic acid) (poly (L-lactic acid, poly (dextran lactic acid, poly (lactic acid)) poly (glycolic acid) (combination of the two, poly (lactic acid and poly (glycolic acid))	Constructed by electrospinning, self-assembly, phase separation, *etc.*
Advantages	It comes from organism, with high biocompatibility, compliance and low immune rejection	Adjustable microstructure, surface morphology, mechanical properties and degradation rate Degradable cycle, eventually converted to water and carbon dioxide, easy to process	Provide a controllable environment for cell proliferation and directional differentiation, and significantly improve antithrombotic function
Disadvantages	Lack of ability to control extracellular matrix deposition and construction, with the potential risk of transmission of related animal origin pathogens	The synthesis technology is high and expensive, lacking special biological signal or functional group	The synthesis technology is high and expensive, lacking special biological signal or functional group

Natural biological scaffolds are derived from organisms and can be divided into macromolecular unstructured materials and acellular matrix scaffolds. The former includes chitosan, alginate, collagen, gelatin, and hyaluronic acid while the latter includes acellular dermal matrix, acellular small intestinal submucosa, SIS, and acellular vascular matrix. This kind of biomaterial has a strong affinity toward cells. The biomaterial itself contains cytokines and biological signalling molecules that can promote the growth and differentiation of cells similar to *in vivo* tissues, which is conducive to promoting the combination of cells and scaffolds to promote wound repair and healing, especially acellular matrix scaffolds, which only remove cells from the tissue. The strength and structure of acellular matrix scaffolds are almost the same as that of pre-acellular tissues because the spatial conformation of their structural proteins remains relatively intact. However, at the same time, there are some deficiencies in natural biomaterial scaffolds. For example, natural biomaterial lacks the ability to control the deposition and construction of extracellular matrix, and because it is derived from animal tissues, there are potential risks in association with pathogen transmission [4].

For the application of natural bioscaffold materials, studies have shown that the porous scaffolds made up by cross-linking chitosan along with gelatin by freeze-drying technology have a three-dimensional structure which provides sufficient growth space for fibroblasts [8]. Uniform distribution in the C2G5 type scaffold material, deep into the inner part of the three-dimensional scaffold, and on other scaffold materials is limited to the growth of the scaffold surface. In addition, collagen is widely used in natural bioscaffold materials [9, 10], which has good biocompatibility and is easy to obtain. Extracellular proteins such as collagen have a nanofiber structure, and their diameter is between 50 and 500 nm. Using collagen can enhance cell adhesion and promote cell proliferation and differentiation on the surface of the scaffold. The collagen nanofiber biomimetic scaffold consists of degradable polymer nanofibers that can be achieved by electrospinning or self-assembly. Chan *et al.* [11] found that porous three-dimensional collagen scaffolds can effectively support the formation of capillaries *in vitro* and promote the vascularization of tissues after implantation *in vivo*. In their experiment, the number of fat stem cells in the experimental group with collagen scaffolds was significantly greater than that in the control group without collagen scaffolds.

It can be proven that using three-dimensional collagen scaffold is an effective way to promote angiogenesis and three-dimensional scaffold.

Studies have shown that acellular matrix scaffold materials also have broad application prospects. For example, acellular vascular matrix is a natural structure in which blood vessels are specifically treated to remove only cellular components but does not damage the extracellular matrix. The amino acid residue sequence in the matrix can be the integrin receptor, and recognition of the cell membrane promotes the compatibility of the cell with the matrix, thereby promoting the proliferation and differentiation of the cell. Since the scaffold material has strong affinity toward cells, it can provide a cell with an approximate environment *in vivo* for cell growth, proliferation, and differentiation. Meanwhile, the immunological repellency, biocompatibility, and biomechanical properties of the outer matrix are higher than those of synthetic materials, and it has received increasing attention in the field of tissue engineering vascular construction. In addition, the collagen fiber and the elastic fiber components in the extracellular matrix have the function of maintaining the mechanical strength of the blood vessel. Therefore, like the normal vascular tissue, the acellular vascular matrix also has good mechanical properties and can bear a certain amount of load when implanted in the animal body, with blood flow impact. In the study of acellular matrix scaffold materials, SIS is also considered an alternative material for tissue-engineered blood vessels. SIS is composed of fibronectin, proteoglycan, collagen, glycoprotein, *etc.*, and has been shown to have sufficient burst pressure, far exceeding In vein grafts, studies have shown that the implantation of SIS into animal arteries has good research prospects [7, 12].

Synthetic materials can be divided into non-degradable polymer materials and degradable polymer materials. Non-degradable polymer materials cannot be degraded in the body. These materials can be classified into polyester and expanded polytetrafluoroethylene. In a prior study, fibrin was pre-applied on expanded polytetrafluoroethylene to inhibit fibrin glue to achieve endothelialization *in vitro*. When applied to femoral-iliac artery shunt, the openness of small-caliber vascular grafting was improved [13, 14]. Despite this, the defects of polymer materials such as expanded polytetrafluoroethylene and polyester that cannot be degraded or modified, have limited their potential in vascular endothelialization applications. Synthetic degradable polymer materials are synthesized by chemical methods. Commonly used degradable polymer materials mainly include poly-β-hydroxybutyric acid, polyamino acid, polycarbonate, polyurethane, polylactic acid (there are three stereo configurations namely, poly-L-lactic acid, poly-dextran lactic acid, and poly- lactic acid), polyglycolic acid (and a combination of the two, polylactic acid and polyglycolic acid), poly-ε-caprolactone, and the like. The microstructure, surface morphology, mechanical properties, and degradable rate of synthetic biodegradable polymer materials can be regulated, and there will be no pathogen infection, no tissue and antigen reaction [15] after implantation into the body by polylactic acid degradable polymer organisms. The material has good biocompatibility, non-toxic reaction, and can be gradually degraded in the body, that is, it has a tunable biodegradation cycle, and is finally converted into water and carbon dioxide; it is easy to process, and is an ideal tissue engineering vascular scaffold material [16]. Although biodegradable macromolecule biomaterials such as polylactic acid and polyglycolic acid have many advantages, due to the high technology of material synthesis process and the high price, coupled with the lack of special biological signals or functional groups [31 - 33], the affinity between the material surface and cells is insufficient, which limits the application of the material itself. However, the compatibility and bioactivity of these synthetic biomaterials with seed cells are key factors [17] to improving this situation.

Among many synthetic degradable polymer materials, composite materials have become the focus of research due to their adjustable degradation rate and good mechanical properties. These include the combination of two natural biomaterials, the combination of natural biomaterials and synthetic biodegradable polymer materials, and the combination of two synthetic biodegradable polymer materials. At present, PLGA has become the most widely used material in composite materials. Yu *et al.* [36] showed that accelerated and controlled vascularization on scaffolds remains a major limitation in tissue engineering applications. In this experiment, 2-N, 6-O sulfated chitosan was coated on PLGA scaffolds as a carrier of vascular endothelial growth factor, or subsequently promoted the formation of blood vessels *in vitro*. The results showed that PLGA scaffolds modified with 2-N, 6-O sulfated chitosan had the advantage of slowly releasing vascular endothelial growth factor. In addition, the viability and adhesion of human umbilical vein endothelial cells (HUVECs) were good after they were inoculated into the reconstructed scaffolds. The release of vascular endothelial growth factor can promote the formation of capillaries. These results fully indicate that the enhancement of angiogenesis is due to the modification of PLGA scaffolds.

Taking PLGA as an example, PLAG-cross-linked chitosan electrospun nano-bi--composites provide better cell adhesion and proliferation rate for human adipose-derived stem cells [18]. Duan *et al.* [19] introduced vascular endothelial growth factor gene into mesenchymal stem cells by adenovirus transfection. The expression of exogenous gene was determined by enzyme-linked immunosorbent assay. Then the transfected cells were inoculated into PLGA/tricalcium phosphate (TCP) scaffolds modified by collagen type I. Composite artificial bone was constructed *in vitro* and finally vascular and osteogenesis were achieved *in vivo*. The results showed that cross-linking collagen type I with porous PLGA scaffolds could improve the biocompatibility of mesenchymal stem cells transfected by adenovirus. In addition, *in vivo* experiments showed that a large number of neovascularization occurred 8 weeks after implantation. This study shows that vascular endothelial growth factor gene-transfected mesenchymal stem cells combined with PLGA/TCP can be used in bone *in vivo*. Yao *et al.* [20] showed that thrombosis and occlusion of the lumen were the main reasons for the failure of small-caliber vascular construction. The study used electrospinning technology to combine poly(epsilon-caprolactone) synthetic biomaterials with natural biomaterial chitosan. The surface of the synthetic biomaterials was fixed with heparin to construct small-caliber vessels. The results showed that the synthetic scaffolds could not only effectively reduce blood flow. Platelet adhesion can prolong the coagulation time, promote the proliferation of human vein endothelial cells, prevent the excessive growth of vascular smooth muscle cells, and finally successfully construct small-caliber vascular substitutes. Recently, in order to overcome some of the shortcomings of synthetic biodegradable materials, some researchers combined artificial synthetic polymer materials with natural acellular vascular matrix to produce tissue engineered vascular substitutes with mixed components [21]. Under vacuum lyophilization conditions, the donor's arterial vascular wall cells are removed using a detergent combined with a dehydrating agent, and then the poly-ε-caprolactone material is immobilized on the surface of the natural acellular vascular matrix scaffold material by electrospinning. Then enhance the strength properties of the acellular matrix and coat the surface of the hybrid scaffold material with heparin before being transplanted into the animal body. The biomechanical test results showed that the electrospun poly-ε-caprolactone scaffold material can significantly enhance the natural acellular vascular matrix scaffold material. After transplantation of rat models, biomechanical properties of electrospun poly-ε-caprolactone scaffolds can effectively prevent vasodilation and aneurysm and can effectively reduce inflammatory cell infiltration. The good histocompatibility and biomechanical properties of electrospun polyε-caprolactone scaffolds provide favorable technical support for the construction of small-bore vascular substitutes.

Recent studies have shown that nanomaterials or nanotechnology-treated materials are also widely used as tissue engineered vascular scaffold materials. Nanomaterials or nanotechnology-treated materials can provide a controlled environment for cell proliferation and directed differentiation, Nanovascular stent material can regulate the expression of endothelial cell phenotype [22]. In addition, such materials are constructed using electrospinning, self-assembly, phase separation, *etc.*, and the application of electrospinning technology is most widely used in the construction of nanovascular stent materials [23, 24]. Ahn *et al.* [25] and Tillman *et al.* [26] used electrospinning technology to polymerize ε-caprolactone. Type I collagen is used in the manufacture of vascular scaffold materials to mimic the structural and biomechanical properties of natural blood vessels. In a research, it was found that this biodegradable tubular scaffold material can maintain vascular physiological functions for up to one month *in vitro*, and after transplantation into animals, the scaffold material maintains its structural integrity. The use of nanotechnology to construct tissue-engineered vascular stent materials can significantly improve antithrombotic function, but how to maintain antithrombotic function in the body for a long time has not been solved. Mun *et al.* [27] implanted vascular smooth muscle cells on a new polylactic acid poly-ε-caprolactone composite vascular scaffold prepared by electrospinning technology to construct a small-caliber blood vessel graft substitute with tensile strength, tensile strain and elastic modulus. Both are similar to natural blood vessels, indicating that nanovascular stent materials constructed by electrospinning technology have opened up new avenues for vascular repair and regeneration.

2.3. Topological Structure of Scaffold Material

The topological morphology of biomaterials is one of the important factors affecting cell behavior. In recent years, with the development of nanobiology and medical technology, research has found that cells can "induct" contact to induce the nano-topological structure of substrate materials and produce a strong response. The topological structure and spatial topological structure of the surface of the scaffold material, especially the woven structure, can affect the process of vascularization of related treatment areas by affecting cell adhesion, proliferation, directional growth, and biological activity.

The porous structure itself can directly determine cell growth and nutrient transfer efficiency. It is recognized that several important parameters such as pore size, porosity, pore connectivity, and surface area, can be used to describe the porous structure. For an ideal biological scaffold material, porous structure is the key to cell growth and nutrient transport. According to the literature, pores have traditionally been considered a prerequisite for bone regeneration, with preferred diameters ranging from 200–350 μm. Depending on the specific situation, different cell behaviors apply to different pore sizes. For example, large pore size will reduce cell seeding efficiency, but high cell viability seems to be related to larger pore size. Amini *et al.* [28] showed that a larger pore size can improve the oxygen diffusion capacity inside the material and increase cell viability. Another study showed that for the growth of blood vessels, more than 400 μm pores were preferred [29]. Therefore, the optimal pore size needs to be analyzed in specific circumstances, and most scholars currently do not recommend the preparation of unimodal scaffold materials with small holes for bone or blood vessel formation.

Studies have shown that pore size less than 100 μm in diameter may prevent the transfer of oxygen and nutrients to the center of the scaffold [30], because cells may accumulate on the surface of the scaffold and block the pores. Second, as mentioned before, for a biomimetic gradient porous structure, both macropores and micropores are conducive to angiogenesis [31]. In a study by Salerno *et al.* [32], two types of PCL scaffolds were prepared. One POS evaluates the single mode pore distribution (with an average pore size of 325 μm) and the other with a bimodal pore distribution (with an average pore size of 38312 μm). These two scaffolds are superior to the traditional scaffolds in terms of cell inoculation efficiency, cell survival rate, and cell proliferation ability. In addition, the existence of micropores is conducive to protein adsorption, and improves porosity and pore connectivity. The stiffness of the matrix also needs attention. Biophysical stimulation is very important in the process of angiogenesis. The shape of endothelial cells affects angiogenesis. Research shows that the reason why angiogenesis requires matrix stiffness may be attributed to vascular endothelial growth factor and so on, but there is an upper limit stiffness [33]. Wu *et al.* [34] found that good vascularization could be observed on adaptive gel (11 to 36 thousand PPA). However, because of the different stiffness required by arterial and venous vessels, researchers are working hard to produce abnormal transduction models that mimic the rigid regulation of extracellular matrix and differentiate endothelial stem cells. These studies provide new concepts for biomaterial design.

3. TISSUE ENGINEERING STRATEGY FOR INDUCING ANGIOGENESIS USING BIOLOGICAL SCAFFOLDS

3.1. Scaffold-Based Delivery System

The use of scaffold materials as a bridge to deliver substances to cells is the most direct use of scaffold materials. Many nanomaterials are a popular choice for scaffold materials because of their good biosafety and editability, and can carry many nucleic acid related drugs such as Micro RNA, siRNA, LncRNA, and nucleic acid aptamers. These drugs all change the progression of vascularization from within the cell.

Zhao *et al.* [35] found that tetrahedral DNA nanostructure can enter into endothelial cells (ECs) and promote EC proliferation, migration, tube formation, and the expression of angiogenic growth factors, which was accompanied by activation of the Notch signaling pathway. Xie *et al.* [36] combines a nucleic acid aptamer called pegaptanib with a tetrahedral DNA nanostructure. The result shows that pegaptanib-loaded TDNs could effectively enhance the ability of pegaptanib to inhibit proliferation, migration, and tube formation of HUVECs induced by VEGF. Na-Kyung Ryoo *et al.* [37] invented a kind of nanoball loaded with siRNA to interfere with the production of VEGF. The anti-VEGF effect of intravitreally-injected siVEGF lasted for at 2 weeks showing high targeting efficiency. Microvessel density and maturity were superior to those of the VEGF or bFGF group alone. Zeng *et al.* [38] injected a lentiviral vector transfected with miRNA-210 into the mouse brain. Compared with the untransfected group, the proliferation rate of the endothelial cells and the number of newly formed microvessels in the experimental group increased significantly.

In addition to drugs that deliver nucleic acid analogs, scaffolding materials can also be loaded with small molecule drugs and polypeptides to modulate angiogenesis. Jiang *et al.* [39] used a bladder acellular matrix coated with VEGF and bFGF nanospheres for bladder defect repair. The results showed that the experimental group can significantly improve the contraction of the new bladder tissue, and the arrangement of urothelial cells and muscle cells in the new tissue. Hyun Woong Kim *et al.* [40] mixed pigment epithelium-derived factor (PEDF)–derived 34-mer peptide with type I collagen to improve the effect of PEDF-34, and the result showed that collagen broadened the effective dose range of PEDF-34 in the tube formation assay by >250 times. Yin *et al.* [41] has created a novel multi-target inhibitor that contains multiple binding domains for VEGF and PDGF receptors to regulate angiogenesis.

Although this method of delivering drugs to the interior of cells to regulate angiogenesis is somewhat innovative, the limitations are obvious. Angiogenesis is not the result of the action of a cell or a cytokine. This highly targeted approach does not fully regulate angiogenesis. At the same time, when using nanomaterials to deliver drugs, the speed needs to be finely regulated to ensure that the drug concentration is at an ideal level. The consequence of the carrier remaining after delivery is another problem that needs to be studied. Future research should focus on combining several methods to achieve better results.

3.2. Special Treatment of Scaffold Material to Promote Vascularization

After special treatment, the scaffold material can stimulate the secretion of angiogenic factors from surrounding tissues or seed cells and recruit endothelial cells to promote vascularization. At present, there are two main directions for exploring vascularized scaffold materials. One is that synthetic polymer materials promote vascularization by virtue of their unique physical and chemical properties, and the other is improved processing of natural materials, making them a better vascularization scaffold material.

A considerable amount of material has been invented for synthetic scaffold materials. Zhao *et al.* [42] found that complex 3.0% CuO borate bioactive glass microfibers can promote endothelial cell migration, assemble blood vessels and secrete VEGF, and upregulate the expression levels of angiogenesis-related genes in fibroblasts. Qin *et al.* [43] co-cultured a 1% strontium calcium phosphate scaffold with human dental pulp cells. Using calcium polyphosphate and hydroxyapatite scaffolds as controls, it was found that the experimental group not only promoted the proliferation of dental pulp cells, but also significantly stimulated the cells to secrete more VEGF and bFGF, suggesting that the composite 1% strontium calcium phosphate scaffold has certain potential in inducing vascularization of dental tissues. Quinlan *et al.* [44] used bioactive glass/collagen mucopolysaccharide scaffolds for bone tissue engineering. *in vitro* experiments showed that bioactive glass particles with a diameter of 100 μm significantly promoted the secretion of VEGF by endothelial cells in the scaffold and promoted the formation of endothelial cells; the scaffold promotes the proliferation and osteogenesis of osteoblasts. Koc *et al.* [45] constructed a tissue-engineered bone tissue with a chitosan/hydroxyapatite composite scaffold, which also achieved good results.

At the same time, with the maturity of 3D printing technology, 3D printing materials are gradually favored because of the high degree of editability of their shapes, and increasingly more 3D printed vascular stent materials have emerged. Li *et al.* [46] used 3D printing technology to construct a calcium phosphate cement composite mesoporous silica scaffold with a specific pore structure to optimize the release of silicon ions, which promoted the growth of peripheral blood vessels early in the implant. Bertassoni *et al.* [47] constructed a scaffold containing microvascular channels using a methacrylate gel, polyethylene glycol ester, and dimethacrylic acid based on agarose gel. The microvascular channel can complete cell material exchange and provide cell growth space, supporting the adhesion and proliferation of endothelial cells. Lin *et al.* [48] invented graphene-contained calcium silicate (CS)/polycaprolactone (PCL) scaffold, which possess the dual bioactivities of reaching osteogenesis and vascularization for bone regeneration.

Despite the significant advances in synthetic vascularized stent materials, due to current manufacturing techniques and limitations of the materials themselves, such as inflammatory reactions and toxicity of degradation products, synthetic polymer scaffolds are still insufficiently interacting with seed cells or organisms. It is impossible to construct a mature and stable vascular network and further optimization is needed. In contrast, natural biomaterials have close-in side chains and good biocompatibility, and are easily integrated with host vascular networks, making them another research hotspot.

Wang *et al.* [49] modified the zein porous scaffold with fatty acids. The *in vivo* experiments showed that the biocompatibility was better. Compared with the untreated group, a denser microvascular network was formed in the scaffold, and the degree of fibrosis was slight. Scientists have found that using a type I bovine collagen can make a collagen scaffold with an 80 μm pore size and surrounded it around the femoral artery [50, 51]. The results showed that the scaffold had better biocompatibility, improved the survival rate of the transplanted cells, and the stent was filled with femoral artery and branched small blood vessels. Studies have found that the composition ratio, hardness, and mechanical stimulation of fibrin and type I collagen complexes can affect the construction of vascular networks.

In addition, the extracellular matrix after decellularization has better biocompatibility as a vascular scaffold material. Gálvez-Montón *et al.* [52] covered the acellular matrix of the pericardial source with an infarcted myocardium. After 30 days, microvascular network and nerve fiber regeneration were observed in the acellular matrix, and the left ventricular ejection fraction and cardiac output were significantly improved. The infarct size was significantly reduced. Lyyanki *et al.* [53] used the acellular dermal matrix of composite adipose stem cells to repair the abdominal wall of rats, and the composite adipose stem cells significantly increased microvessel density and mechanical strength at 4 weeks after surgery. Acellular matrix can exert its unique advantages in the field of whole organ tissue engineering and deserves further study.

3.3. Scaffold-Based Cell Complex

According to the foregoing, angiogenesis is caused by a series of events, and the formation of vascular networks requires precise regulation of many pro-angiogenic factors and cells. Therefore, it is a new idea to use scaffold materials to change the number of cells and cytokines in the internal environment, thereby regulating the process of vascularization.

Angiogenesis-related cells include: 1) directly related cells, including endothelial cells, pericytes, and vascular smooth muscle cells; 2) indirectly related cells, including endothelial progenitor cells (EPC), mesenchymal stem cells (MSCs), and induced pluripotent stem cells (iPSC), which promote vascularization by paracrine angiogenic factors or by direct differentiation into endothelial cells.

Sun *et al.* [54] found that co-culture of outgrowth endothelial cells with human mesenchymal stem cells in silk fibroin hydrogels promotes angiogenesis. Buitinga *et al.* [55] implanted islet cell complex hBMSC and umbilical vein endothelial cells subcutaneously into nude mice. Compared with that in the control group without endothelial cells, the tissue microvessel density was higher, which may be due to the interaction of hBMSC and endothelial cells, resulting in a large amount of VEGF. Angiogenic factors such as bFGF are secreted. Zigdon-Giladi *et al.* [56] implanted a β-calcium phosphate material of human peripheral blood EPC into a nude mouse model of craniofacial regeneration, and found that the vascular density of the newly formed bone tissue was 7.5 times higher than that of the simple β-calcium phosphate material group. The microvascular fragment does not only releases angiogenic factor (AF), but also contains many adipose stem cells and EPC, and the adipose stem cells have stronger differentiation and proangiogenic ability than the adipose stem cells obtained by the conventional method. The microvascular segment has a normal vascular morphological structure, contains a central lumen, surrounded by epithelial cells and parietal cells, and only needs to be interconnected to form a microvascular network, which greatly shortens the time required for vascularization. Studies have applied microvascular fragments to muscle and bone tissue engineering and have achieved better vascularization effects [57]. Engineering tissue vascularization requires a variety of cell involvement. Recent studies have found that pericytes and vascular smooth muscle cells are important for the stabilization and functionalization of neovascularization, and that there are morphological and functional aspects of endothelial cells in different arteries, veins, or different tissues and organs. Differences should be noted in subsequent studies.

3.4. Prevascularization to Promote Vascularization of Engineered Tissue

Prevascularization is based on the above three vascularization strategies, where there is a specific phase of *in vitro* or *in vivo* incubation of the vascular network prior to transplantation of the target site. Prevascularization strategies include both *in vitro* and *in vivo* approaches. The *in vitro* route mainly refers to *in vitro* culture of angiogenesis-related cells, and the formation of a microvascular network for *in vivo* transplantation. *In vivo* pathway refers to the implantation of the engineered tissue into the vascular-rich site of the host, allowing the surrounding blood vessels to grow into the tissue and then transplanting the tissue to the target site.

Prior to *in vivo* transplantation, angiogenesis-related cells such as endothelial cells are implanted into the scaffold material, and then the microvascular network is incubated in a specific *in vitro* environment to shorten the time of formation of the vascular network after implantation in the tissue. The most widely used method of *in vitro* prevascularization is to inoculate angiogenic cells onto a scaffold that is of synthetic origin or consists of a natural acellular matrix [58, 59]. Sakaguchi *et al.* [60] repeatedly superimposed three layers of cardiomyocyte-endothelial cell co-cultured cell sheets on a microchannel collagen gel perfused with medium, wherein the endothelial cells were assembled into microvessels and migrated into the collagen gel, and remained in the collagen gel. The microchannels establish a connection that is visible to the naked eye by the simultaneous contraction of engineered myocardial tissue constructed by stacking cell sheets. However, endothelial cells have the following major drawbacks: they cannot be easily harvested in large quantities under clinical conditions, and they do not exhibit high proliferative activity during culture. In addition, endothelial cells derived from different types of blood vessels and different organ tissues differ significantly in their homeostasis, molecular permeability, vascular tone, immune tolerance, and angiogenic potential. Therefore, various methods that can replace endothelial cells have been proposed. Endothelial progenitor cells (EPCs) have been recognized as a promising alternative to tissue engineering methods, and physicians can harvest these cells with minimal invasiveness from bone marrow or peripheral blood [61, 62]. In addition to advanced EPC, other cell types are also suitable for angiogenesis of tissue constructs. These include pluripotent mesenchymal stem cells from bone marrow (MSCs) [63, 64] or adipose tissue [65, 66], amniotic fluid-derived stem cells [67] *etc.*

In the past, prevascularization has often been performed by placing the stent in a blood-rich and easy-to-operate position, such as a subcutaneous or muscle "pocket." Although the above method can vascularize the constructed tissue, it still takes a long time to communicate with the host blood vessel after transplantation to the target site, which may cause ischemic death. Therefore, new *in vivo* prevascularization strategies have emerged, such as arteriovenous circuits, tissue flap techniques with axial blood vessels, *etc.* Tatara *et al.* [68]. fixed the polymethyl methacrylate chamber on the rib membrane and filled the bone into the room. While the bone tissue was renewed in the small chamber, the periosteal vascular network grew into the new bone tissue by budding, thereby obtaining vascularized bone organization. Because the vascular network of the rib membrane and the adjacent intercostal arteries and veins communicate with each other, a free bone tissue flap with the intercostal arteriovenous axis as the axis is constructed, and the mandibular defect is successfully repaired. The above strategy allows the engineered blood vessels to conform to the blood vessels near the target, and blood flow perfusion can be formed immediately after transplantation. Kaempfen *et al.* [69]. Implanted BMSCs with acellular bone matrix to repair rabbit segmental tibiofibular defect models in three ways: 1) direct orthotopic transplantation; 2) grafting a vascular branch of the radial artery to the defect; 3) was wrapped with a muscle flap of the radial artery as the axis and incubated for 6 weeks in the body, and then transplanted to the defect. It was found that the vascular density of the engineered bone tissue constructed by the third method was significantly higher than the other two.

Extensive *in vitro* and *in situ* experiments have developed methods in the past few decades. Successful transfer of these methods to clinical routines is one of the major challenges in tissue engineering. Despite impressive progress in the field, additional preclinical and clinical analyses are needed to further optimize the efficiency and safety of different concepts.

4. BIOLOGICAL CONSIDERATION OF SCAFFOLD MATERIALS IN INDUCING ANGIOGENESIS

4.1. The Outcome of Scaffold Materials in the Process of Vascularization

According to the foregoing, scaffold material exerts different effects on angiogenesis through its own special physical and chemical properties or the drugs it carries. However, there is still no research on the products after the decomposition of the scaffold material and the effects of the products on the cells and the internal environment. Understanding these issues will help researchers design a more secured scaffold material, which is more conducive to the stent material to play a stronger role. Therefore, next we will discuss the related research on the final fate of the scaffold material in degradation.

An excellent degradation process has the following advantages for angiogenesis caused by scaffold materials: 1. Provides space for cell infiltration and new tissue. 2. Separates from the drug being loaded to allow the drug to better bind to the target. 3. Avoids surgery to remove the scaffold and reduce patient damage. In addition to affecting the function of the scaffold material, a poor degradation process also has a reaction at the above three time points. Therefore, understanding the mechanism of scaffold material degradation can help us design more accurate pre-programmed degradable scaffold materials. There are usually two types of scaffold degradation: bulk erosion and surface erosion. Surface erosion of the polymeric device leads to a gradual decrease in the size as degradation proceeds from the surface whereas bulk erosion does not change the size of the polymeric device; rather it weakens the structure as degradation occurs throughout the polymeric device. Different hydrolysis methods are mainly related to two factors: water diffusion rate and hydrolysis rate. For materials where water can easily diffuse into the interior of the polymer, it is often the case of bulk erosion. When the degradation rate of the polymer itself exceeds the rate of water diffusion, the corrosion process is dominated by surface erosion [70]. Of course, these two degradation processes are not independent. In the internal environment, these two forms of degradation often act on the same scaffold material due to various physical and chemical properties.

4.2. Modes of Degradation

Macroscopically, the first step in the degradation of scaffold materials is the corrosion of the material. However, microscopically, the degradation of materials is still done by various chemical reactions. According to the nature and composition of different materials, the degradation of materials mainly includes three modes: Hydrolytic degradation, Enzymatic degradation, and Stimuli-response degradation. Most of the absorbable materials, including most natural high molecular compounds and some synthetic absorbable materials (including poly (glycolic acid) (PGA), poly (lactic acid) (PLA), poly (lactic-co-glycolic acid) (PLGA)) [71 - 73], are decomposed into water and carbohydrates through these three degradation pathways, and can be absorbed by the body. Next, we will explain the specific process of these three degradation modes one by one (Fig. **1**).

Hydrolytic degradation refers to the interaction of a living organism with a hydrolysable, labile bond in the backbone of a degradable material. Common chemical groups of this type are: esters, orthoesters, carbonates, amides, anhydrides, and carbamates [74]. Briefly, hydrolytic degradation begins with the diffusion of water, which in turn causes the random bonds of the amorphous regions of the compound to break. Hydrolysis of sensitive bonds exposes the acid end and produces oligomers or water soluble monomers [75, 76]. This process changes the mechanical and physical properties of the scaffold material in a gradual repetition, and ultimately, the constant cleavage and mass loss of covalent bonds in the backbone due to too much newly formed chain ends.

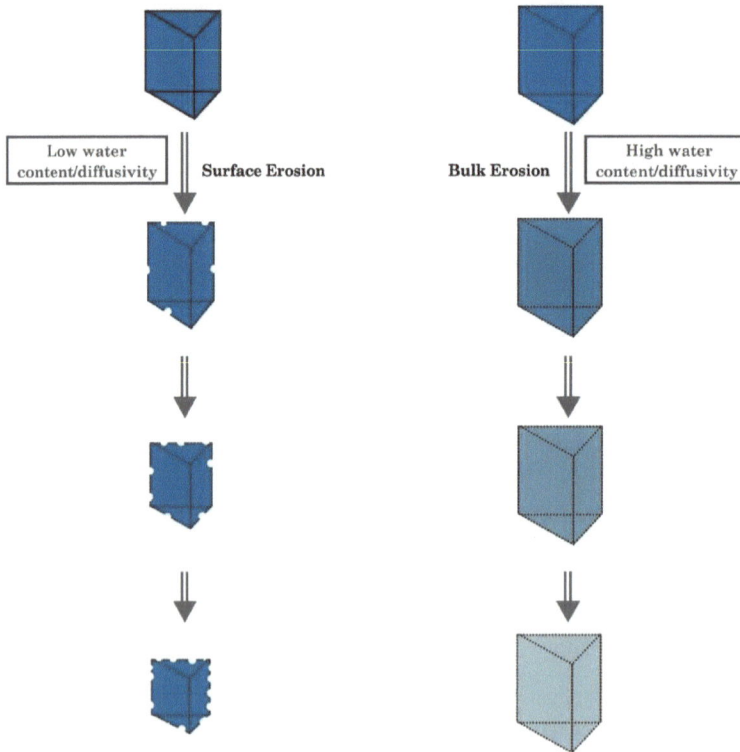

Fig. (1). Schematic diagram of the degradation modes of two different scaffold materials.

The second mode of degradation is enzymatic degradation. The main function of the enzyme is to reduce the activation of the reaction so that the degradation can be carried out under relatively mild conditions [77]. For the pores of the material, the enzyme is too bulky to enter the center of the material. Therefore, the enzymatic degradation reaction is generally a surface erosion reaction.

Table 2. The factors influencing the degradation of scaffolds.

Parameters	Results
Shapes	Spiral wound sheet can enhance cell migration to regenerate bone efficiently due to macrochannel effect
Sizes	Increasing the thickness of PLGA scaffold from 0.05 mm strut size did not affect degradation behavior, while the degrading rate was slowed at the thinnest 0.002 mm strut size
Porosity/pore size	PLLA scaffold with pore size at 0.82 mm (PLLA-L), 0.55 mm (PLLA-M), 0.28 mm (PLLA-S)and solid cylinder (PLLA-C)was applied. PLLA-S showed the least mass loss at 7, 14 week with no significant difference. In 21 week the statistical difference of mass loss between PLLA-S and other groups was observed
Monomer structure	Methyl groups can hinder water penetration into PLA based scaffolds, resulting in a slower degradation rate than other otherpoly(α-hydroxy esters). Classes of hydrolysable chemical bonds and their degradation rate order based on corresponding half-lives: polyamides<polyesters<poly(ortho esters) < polyanhydrides
Crystallinity/glass transition temperature (Tg)	Adding oligomer to PLGA scaffolds with higher content results in lower crystallinity and Tg. Using molecular weight change as the test for degradation profile, it is showed that higher crystallinity and Tg contribute to slower water absorption and scaffold degradation
Tacticity	In regard of the two types of stereoisomers of PLA, D-lactic acid polymer degrades faster than L-lactic acid polymers
Scaffold inner property	Hydrolytic degradation is a major process of scaffold collapse. Hydrophobic materials suffers slower degradation due to the less water permeability
Loading direction	Scaffold alignment of filaments influence degradation rate and porosity as well, 0/30 lay-down pattern degrades slower than 0/90, and possesses smaller porosity
Proportion of polymer	Electrospun SF/P(LLA-CL) blends hinder the movement of molecular chains, hence slow the degradation. Glycolic acid content contributes to the different degradation of PDLLA and PLGA components. Boron nitride nanotubes can decrease degradation rate of gelati-glucose scaffold *via* enhancing mechanical strength. Carbonated apatite ratio has a strong influence on limiting PLGA degradation
Treatment	Plasma treatment can damage the integrity of scaffold, leading to faster degradation. Increased crosslinking density and deposition layer-by-layer limit water filtration speed and degradation as well
Drug loading	Degradation rate and scaffold collapse speed are in consistent with drug loading in a dose dependent manner, due to the dispersion of drug particles from the scaffold and high degradability *in vivo*
Physiological environment	Alkaline and acidic pH accelerate degradation rate compared with polymers immersed in solution with neutral pH. Scaffolds usually degrade faster *in vivo* than *in vitro*, and the degradation rate varies in different experiment animals

As mentioned earlier, scaffold degradation is a "double-edged sword" for tissue regeneration and immune response. On the one hand, the integrity of the scaffold provides the surface for osteoclasts and osteoblasts to attach, grow, and excrete osteoblasts. The degradation of the scaffold also provides space for newly formed fillers. A scaffold with a slow degradation rate will inhibit the regeneration effect and stimulate the host foreign body reaction. The collapse of the stent may cause a large number of by-products in a short time, which is harmful to the repair of the implanted area.

How to control the degradation of the stent at the ideal speed, how to control the interaction between the stent and the internal environment, and how to achieve a better balance between the stability and degradability of the stent are all problems that are worthy of our next experiment To continue to explore the direction. Therefore, in addition to understanding the basic processes and pathways of scaffold degradation, we should also understand the various environmental factors in the degradation of scaffolds, and the impact of physical and chemical factors on this process.

The enzyme first combines with the material on the surface to form an enzyme-substrate complex, which then undergoes a catalytic reaction at the interface of the complex, resulting in a rapid decrease in molecular weight and mass loss [78]. At the same time, in the hydrolysis reaction, the porosity of the material becomes large due to hydrolysis, which promotes the progress of the enzymatic reaction to a certain extent. When the hydrolysis reaction is completed, the body inside the stent is exposed, and the attachment of the enzyme is further recruited [79].

The last form of degradation is to stimulate the associated scaffold dissolution process. The main process is that the scaffold network structure is cleaved by external triggers to cause swelling or sol-gel degradation behavior [80]. A representative of this material is a smart hydrogel. Because of the artificial control of the timing of material lysis, the appearance of this material provides a more convenient method of modification for drug delivery [81]. The most common stimuli for polymer hydrogels are pH, light, and redox reactions. Chemical bonds including acetals/ketals, orthocenters, imines, hydrazides, and dimethyl maleate are unstable in acidic environments. The normal pH of the human body is between 6.5 and 7.2, and the pH of the human body varies in different areas. For example, in areas where revascularization is required after trauma, the pH may decrease due to damage to the blood supply and inflammatory response [82]. Therefore, pH-responsive smart hydrogels provide targeted controlled release behavior for wound areas and slower degradation in normal tissues. Similarly, photosensitive hydrogels can react differently at different wavelengths of light to achieve different goals [83, 84].

4.3. Factors Affecting the Degradation of Scaffold

Due to the diversity and complexity of the degradation pathways of scaffold materials in the body, many factors affect the degradation of scaffold materials. The nature of the internal material of the scaffold itself determines a part of the degradation process, but more is affected by the physical and chemical properties of the internal environment. Next, we will give some examples of experiments on the physicochemical properties and degradation of scaffold materials to illustrate the factors that affect the degradation of scaffold materials. Lara Yildirimer *et al.* [85] found that shape is an important factor affecting the degradation of scaffold materials. They found that spiral wound tablets can enhance cell migration, thereby effectively recruiting more enzymes through the large channel effect to promote degradation of the scaffold material. In terms of size, Reyhaneh *et al.* [86] found that increasing the thickness of PLGA scaffold from 0.05 mm strut size did not affect degradation behavior, while the degrading rate was slowed at the thinnest 0.002 mm strut size. Author links open overlay panel. Carlos *et al.* [87] used a chitosan-based scaffold with molecular weights of 538 ± 48 (CSH), 229 ± 45 (CSM) and 13 ± 3 kDa (CSL) to observe the degradation rate. As the *in vitro* degradation of Mw at 1, 2, 3, and 4 weeks, the mass loss of the scaffold gradually increased. In addition, it has been documented that pretreatment of the scaffold also increases the rate of degradation of the scaffold material. For example, plasma treatment can damage the integrity of the stent, resulting in faster degradation. Increased crosslink density and layer by layer deposition also limits water filtration rate and degradation [88, 89].

Of course, many physical and chemical properties affect the rate of material degradation, and they will be explained in more detail in the next chapter.

5. CUTTING EDGE TREND AND DEVELOPMENT OF SCAFFOLD MATERIALS

5.1. Shortages of Existing Support Materials

Tissue engineered blood vessels have developed rapidly in recent years. So far, no ideal scaffold material has been found. Although natural biomaterials have become a research hotspot, the physical and mechanical properties do not meet the requirements of scaffolds very well, which urgently need the emergence of new materials to better meet the requirements of tissue-based vascular scaffolds, and to achieve the purpose of repair and reconstruction [90 - 92]. In practical application, the co-culture strategy and the influence of angiogenesis molecules should be considered to stimulate the regeneration response of host and promote the germination and strength of blood vessels [93]. In order to optimize the structure and vascular integration of scaffolds, the existing advanced technology of angiogenesis has not been proven to be safe and effective, such as oxygen biomaterials, scaffold free cell sheet technology, the strategy of axial vascularization induced by arteriovenous loop, and so on. The effective perfusion and functionalization of the tube are still challenging. At the same time, it is a hot topic for scientists to design personalized scaffold materials for different organs and successfully promote vascularization in specific areas [94]. In addition, current tissue engineering involves a very broad field, including biological medicine, physics, chemistry, high polymer materials science, and so on. How to effectively take advantage of the multi-disciplinary design and good performance, economic, and practical aspects of each field can largely promote clinical support material from the lab for actual clinical life applications.

5.2. Relevant Frontier Applications

He *et al.* adjusted the pure physical structure, combined with the existing tissue engineering manufacturing technology (fused deposition molding technology and near-field direct writing technology), and developed a microscopic dual-scale scaffold biological 3D printer to design a multi-scale scaffold, greatly improving the biological phase of the original scaffold capacity. The main method is to use a network of coarse fibers (about 100 μm) and a network of ultra-fine fibers (2-3 μm), which can improve mechanical strength and have a good microenvironment for cell growth, which is conducive to cell adhesion and proliferation. The effect of micro-scale on the scaffold material and cell growth microenvironment was verified, and the feasibility and versatility of this 3D printing material in tissue engineering applications were verified.

In tissue engineering, the transport of oxygen and the removal of metabolic waste are important to the success of vascularization. Susmita *et al.* [95] reported that 3D printed magnesium-silicon composite porous tricalcium phosphate scaffolds were used to promote osteogenesis and angiogenesis in a rat model of distal femoral defect. A 30-day *in vitro* ion release in phosphate buffer indicates that these scaffolds have sustained Mg^{2+} and Si^{4+} release, compared with simple tricalcium phosphate scaffolds without magnesium and silicon; magnesium-silicon composite porous 3DP-TCP scaffolds have good bone tissue repair, regeneration, and vascularization capabilities.

Deng *et al.* [96] using the same three-dimensional printing technology, the porous β-tricalcium phosphate scaffold was co-cultured with human umbilical vein endothelial cells (HUVECs) and human bone marrow stromal cells (hBMSCs) to construct a tissue engineering scaffold with fast vascularization and good osteogenesis. The results showed that the 5% cs/beta-tcp scaffold not only stimulated the angiogenesis of co-cultured cells on the matrix gel, but also promoted the formation of micro-columnar structures on the co-cultured cells. Co-implantation provides a new strategy for promoting vascularization and osteogenesis of tissue engineering scaffolds, and provides new ideas for poor angiogenesis in tissue engineering.

In the process of tissue engineering vascularization, and as one of the three major factors of tissue engineering, scaffolds provide good three-dimensional space support for cell adhesion, proliferation, and differentiation, and are the basis for seed cells and growth factors to exert biological effects; thus playing a vital role in the various processes of vascularization. As the beginning, this chapter first introduced several strategies of promoting vascularization of scaffold materials to readers and described the reasons for the use of scaffold materials in vascularization and the necessity of scaffold materials in tissue engineering vascularization. It then focused on how to use the scaffolding materials correctly. The design of scaffold material itself is the key factor for the material to play a role. The scaffold material with ideal biological properties can make the process of vascularization more effective than half the effort. The topological structure of materials and their physical and chemical properties affect the success rate of vascularization to varying degrees. We hope that readers can get the basic knowledge and principles of scaffold design from this chapter. Finally, there are many new ideas for the design of tissue engineering vascularization materials, such as new material processing methods, new ways of cell binding and so on, which have improved the process of vascularization in varying degrees. Scaffold materials have shown attractive prospects and great possibilities in vascularized tissue engineering. Previous studies have found that many materials are related to vascularization, but there arc also many problems to be solved urgently. For example, the balance between degradation rate, biomechanical strength and tissue formation rate of vascular stent materials has not yet been found. Furthermore, new preparation techniques are needed, and design ideas to improve the bracket structure. With the development of materials science and engineering, it is believed that there will be a new type of vascular stent material, which has better performance in all aspects and is more suitable for vascularization and can achieve vascularization more efficiently in the future.

CONSENT FOR PUBLICATION

Not applicable.

CONFLICT OF INTEREST

The authors confirm that the contents of this chapter have no conflict of interest.

ACKNOWLEDGEMENTS

This study was supported by National Key R&D Program of China (2019YFA0110600) and National Natural Science Foundation of China (81970986, 81771125).

REFERENCES

[1]   Langer R, Vacanti JP. Tissue engineering. Science 1993; 260(5110): 920-6.
      [http://dx.doi.org/10.1126/science.8493529] [PMID: 8493529]

[2]   Li S, Sengupta D, Chien S. Vascular tissue engineering: from *in vitro* to *in situ*. Wiley Interdiscip Rev Syst Biol Med 2014; 6(1): 61-76.
      [http://dx.doi.org/10.1002/wsbm.1246] [PMID: 24151038]

[3]   Luong E, Gerecht S. Stem cells and scaffolds for vascularizing engineered tissue constructs. Adv Biochem Eng Biotechnol 2009; 114: 129-72.
      [PMID: 19082932]

[4]   McBane JE, Sharifpoor S, Labow RS, Ruel M, Suuronen EJ, Santerre JP. Tissue engineering a small diameter vessel substitute: engineering constructs with select biomaterials and cells. Curr Vasc Pharmacol 2012; 10(3): 347-60.
      [http://dx.doi.org/10.2174/157016112799959378] [PMID: 22239637]

[5]   Coakley DN, Shaikh FM, O'Sullivan K, *et al.* Comparing the endothelialisation of extracellular matrix bioscaffolds with coated synthetic vascular graft materials. Int J Surg 2016; 25: 31-7.
      [http://dx.doi.org/10.1016/j.ijsu.2015.11.008] [PMID: 26578107]

[6]   Lesman A, Gepstein L, Levenberg S. Cell tri-culture for cardiac vascularization. Methods Mol Biol 2014; 1181(1181): 131-7.
      [http://dx.doi.org/10.1007/978-1-4939-1047-2_12] [PMID: 25070333]

[7]   Singh R, Sarker B, Silva R, *et al.* Evaluation of hydrogel matrices for vessel bioplotting: Vascular cell growth and viability. J Biomed Mater Res A 2016; 104(3): 577-85.
      [http://dx.doi.org/10.1002/jbm.a.35590] [PMID: 26474421]

[8]   Tseng HJ, Tsou TL, Wang HJ, Hsu SH. Characterization of chitosan-gelatin scaffolds for dermal tissue engineering. J Tissue Eng Regen Med 2013; 7(1): 20-31.
      [http://dx.doi.org/10.1002/term.492] [PMID: 22034441]

[9]   Hasan A, Memic A, Annabi N, *et al.* Electrospun scaffolds for tissue engineering of vascular grafts. Acta Biomater 2014; 10(1): 11-25.
      [http://dx.doi.org/10.1016/j.actbio.2013.08.022] [PMID: 23973391]

[10] Lu T, Li Y, Chen T. Techniques for fabrication and construction of three-dimensional scaffolds for tissue engineering. International Journal of Nanomedicine 2013; 8(default): 50-337.
      [http://dx.doi.org/10.2147/IJN.S38635]

[11] Chan EC, Kuo SM, Kong AM, *et al.* Three dimensional collagen scaffold promotes intrinsic vascularisation for tissue engineering applications. PLoS One 2016; 11(2): e0149799.
      [http://dx.doi.org/10.1371/journal.pone.0149799] [PMID: 26900837]

[12] Rong JJ, Sang HF, Qian AM, Meng QY, Zhao TJ, Li XQ. Biocompatibility of porcine small intestinal submucosa and rat endothelial progenitor cells *in vitro*. Int J Clin Exp Pathol 2015; 8(2): 1282-91.
      [PMID: 25973012]

[13] Lumsden AB, Morrissey NJ. Randomized controlled trial comparing the safety and efficacy between the FUSION BIOLINE heparin-coated vascular graft and the standard expanded polytetrafluoroethylene graft for femoropopliteal bypass. J Vasc Surg 2015; 61(3): 703-12.e1.
      [http://dx.doi.org/10.1016/j.jvs.2014.10.008] [PMID: 25720929]

[14]   Rychlik IJ, Davey P, Murphy J, O'Donnell ME. A meta-analysis to compare Dacron versus polytetrafluroethylene grafts for above-knee femoropopliteal artery bypass. J Vasc Surg 2014; 60(2): 506-15.
[http://dx.doi.org/10.1016/j.jvs.2014.05.049] [PMID: 24973288]

[15]   Alvarez-Barreto JF, Linehan SM, Shambaugh RL, Sikavitsas VI. Flow perfusion improves seeding of tissue engineering scaffolds with different architectures. Ann Biomed Eng 2007; 35(3): 429-42.
[http://dx.doi.org/10.1007/s10439-006-9244-z] [PMID: 17216348]

[16]   Hiltunen M, Pelto J, Ellä V, Kellomäki M. Uniform and electrically conductive biopolymer-doped polypyrrole coating for fibrous PLA. J Biomed Mater Res B Appl Biomater 2016; 104(8): 1721-9.
[http://dx.doi.org/10.1002/jbm.b.33514] [PMID: 26348386]

[17]   Lutolf MP, Lauer-Fields JL, Schmoekel HG, *et al.* Synthetic matrix metalloproteinase-sensitive hydrogels for the conduction of tissue regeneration: engineering cell-invasion characteristics. Proc Natl Acad Sci USA 2003; 100(9): 5413-8.
[http://dx.doi.org/10.1073/pnas.0737381100] [PMID: 12686696]

[18]   Razavi S, Karbasi S, Morshed M, Zarkesh Esfahani H, Golozar M, Vaezifar S. Cell Attachment and Proliferation of Human Adipose-Derived Stem Cells on PLGA/Chitosan Electrospun Nano-Biocomposite. Cell J 2015; 17(3): 429-37.
[PMID: 26464814]

[19]   Duan C, Liu J, Yuan Z, *et al.* Adenovirus-mediated transfer of VEGF into marrow stromal cells combined with PLGA/TCP scaffold increases vascularization and promotes bone repair *in vivo*. Arch Med Sci 2014; 10(1): 174-81.
[http://dx.doi.org/10.5114/aoms.2012.30950] [PMID: 24701231]

[20]   Yao Y, Wang J, Cui Y, *et al.* Effect of sustained heparin release from PCL/chitosan hybrid small-diameter vascular grafts on anti-thrombogenic property and endothelialization. Acta Biomater 2014; 10(6): 2739-49.
[http://dx.doi.org/10.1016/j.actbio.2014.02.042] [PMID: 24602806]

[21]   Gong W, Lei D, Li S, *et al.* Hybrid small-diameter vascular grafts: Anti-expansion effect of electrospun poly ε-caprolactone on heparin-coated decellularized matrices. Biomaterials 2016; 76: 359-70.
[http://dx.doi.org/10.1016/j.biomaterials.2015.10.066] [PMID: 26561933]

[22]   Jordan SW, Chaikof EL. Novel thromboresistant materials. J Vasc Surg 2007; 45(6) (Suppl. A): A104-15.
[http://dx.doi.org/10.1016/j.jvs.2007.02.048] [PMID: 17544031]

[23]   Jing X, Mi HY, Salick MR, Cordie TM, Peng XF, Turng LS. Electrospinning thermoplastic polyurethane/graphene oxide scaffolds for small diameter vascular graft applications. Mater Sci Eng C 2015; 49(49): 40-50.
[http://dx.doi.org/10.1016/j.msec.2014.12.060] [PMID: 25686925]

[24]   Ye L, Cao J, Chen L, *et al.* The fabrication of double layer tubular vascular tissue engineering scaffold *via* coaxial electrospinning and its 3D cell coculture. J Biomed Mater Res A 2015; 103(12): 3863-71.
[http://dx.doi.org/10.1002/jbm.a.35531] [PMID: 26123627]

[25]   Ahn H, Ju YM, Takahashi H, *et al.* Engineered small diameter vascular grafts by combining cell sheet engineering and electrospinning technology. Acta Biomater 2015; 16(1): 14-22.
[http://dx.doi.org/10.1016/j.actbio.2015.01.030] [PMID: 25641646]

[26]   Tillman BW, Yazdani SK, Lee SJ, Geary RL, Atala A, Yoo JJ. The *in vivo* stability of electrospun polycaprolactone-collagen scaffolds in vascular reconstruction. Biomaterials 2009; 30(4): 583-8.
[http://dx.doi.org/10.1016/j.biomaterials.2008.10.006] [PMID: 18990437]

[27]   Mun CH, Jung Y, Kim SH, *et al.* Three-dimensional electrospun poly(lactide-co-ε-caprolactone) for small-diameter vascular grafts. Tissue Eng Part A 2012; 18(15-16): 1608-16.

[http://dx.doi.org/10.1089/ten.tea.2011.0695] [PMID: 22462723]

[28]   Amini AR, Adams DJ, Laurencin CT, Nukavarapu SP. Optimally porous and biomechanically compatible scaffolds for large-area bone regeneration. Tissue Eng Part A 2012; 18(13-14): 1376-88.
[http://dx.doi.org/10.1089/ten.tea.2011.0076] [PMID: 22401817]

[29]   Kim K, Dean D, Wallace J, Breithaupt R, Mikos AG, Fisher JP. The influence of stereolithographic scaffold architecture and composition on osteogenic signal expression with rat bone marrow stromal cells. Biomaterials 2011; 32(15): 3750-63.
[http://dx.doi.org/10.1016/j.biomaterials.2011.01.016] [PMID: 21396709]

[30]   Bobbert FSL, Zadpoor AA. Effects of bone substitute architecture and surface properties on cell response, angiogenesis, and structure of new bone. J Mater Chem B.

[31]   Peters EB. Endothelial progenitor cells for the vascularization of engineered tissues. Tissue Eng Part B Rev 2018; 24(1): 1-24.
[http://dx.doi.org/10.1089/ten.teb.2017.0127] [PMID: 28548628]

[32]   Salerno A, Guarnieri D, Iannone M, Zeppetelli S, Netti PA. Effect of micro- and macroporosity of bone tissue three-dimensional-poly(epsilon-caprolactone) scaffold on human mesenchymal stem cells invasion, proliferation, and differentiation *in vitro*. Tissue Eng Part A 2010; 16(8): 2661-73.
[http://dx.doi.org/10.1089/ten.tea.2009.0494] [PMID: 20687813]

[33]   Galarneau A, Guenneau F, Gédéon A, *et al.* Probing interconnectivity in hierarchical microporous/mesoporous materials using adsorption and nuclear magnetic resonance diffusion. J Phys Chem C 2016; 120(3): 5b10129..

[34]   Wu Y, Al-Ameen MA, Ghosh G. Integrated effects of matrix mechanics and vascular endothelial growth factor (VEGF) on capillary sprouting. Ann Biomed Eng 2014; 42(5): 1024-36.
[http://dx.doi.org/10.1007/s10439-014-0987-7] [PMID: 24558074]

[35]   Zhao D, Liu M, Li Q, *et al.* Tetrahedral DNA Nanostructure Promotes Endothelial Cell Proliferation, Migration, and Angiogenesis *via* Notch Signaling Pathway. ACS Appl Mater Interfaces 2018; 10(44): 37911-8.
[http://dx.doi.org/10.1021/acsami.8b16518] [PMID: 30335942]

[36]   Xie X, Zhang Y, Ma W, *et al.* Potent anti-angiogenesis and anti-tumour activity of pegaptanib-loaded tetrahedral DNA nanostructure. Cell Prolif 2019; 52(5): e12662.
[http://dx.doi.org/10.1111/cpr.12662] [PMID: 31364793]

[37]   Ryoo N-K, Lee J, Lee H, *et al.* Therapeutic effects of a novel siRNA-based anti-VEGF (siVEGF) nanoball for the treatment of choroidal neovascularization. Nanoscale 2017; 9(40): 15461-9.
[http://dx.doi.org/10.1039/C7NR03142D] [PMID: 28976519]

[38]   Zeng L, He X, Wang Y, *et al.* MicroRNA-210 overexpression induces angiogenesis and neurogenesis in the normal adult mouse brain. Gene Ther 2014; 21(1): 37-43.
[http://dx.doi.org/10.1038/gt.2013.55] [PMID: 24152581]

[39]   Jiang X, Lin H, Jiang D, *et al.* Co-delivery of VEGF and bFGF *via* a PLGA nanoparticle-modified BAM for effective contracture inhibition of regenerated bladder tissue in rabbits. Sci Rep 2016; 6: 20784.
[http://dx.doi.org/10.1038/srep20784] [PMID: 26854200]

[40]   Kim HW, Roh K-H, Kim SW, *et al.* Type I pig collagen enhances the efficacy of PEDF 34-mer peptide in a mouse model of laser-induced choroidal neovascularization. Graefes Arch Clin Exp Ophthalmol 2019; 257(8): 1709-17.
[http://dx.doi.org/10.1007/s00417-019-04394-z] [PMID: 31222405]

[41]   Yin X, Lin X, Ren X, *et al.* Novel multi-targeted inhibitors suppress ocular neovascularization by regulating unique gene sets. Pharmacol Res 2019; 146: 104277.
[http://dx.doi.org/10.1016/j.phrs.2019.104277] [PMID: 31112749]

[42]   Zhao S, Li L, Wang H, *et al.* Wound dressings composed of copper-doped borate bioactive glass

microfibers stimulate angiogenesis and heal full-thickness skin defects in a rodent model. Biomaterials 2015; 53: 379-91.
[http://dx.doi.org/10.1016/j.biomaterials.2015.02.112] [PMID: 25890736]

[43] Qin H, Yang Z, Li L, *et al.* A promising scaffold with excellent cytocompatibility and pro-angiogenesis action for dental tissue engineering: Strontium-doped calcium polyphosphate. Dent Mater J 2016; 35(2): 241-9.
[http://dx.doi.org/10.4012/dmj.2015-272] [PMID: 27041014]

[44] Quinlan E, Partap S, Azevedo MM, Jell G, Stevens MM, O'Brien FJ. Hypoxia-mimicking bioactive glass/collagen glycosaminoglycan composite scaffolds to enhance angiogenesis and bone repair. Biomaterials 2015; 52: 358-66.
[http://dx.doi.org/10.1016/j.biomaterials.2015.02.006] [PMID: 25818442]

[45] Koç A, Finkenzeller G, Elçin AE, Stark GB, Elçin YM. Evaluation of adenoviral vascular endothelial growth factor-activated chitosan/hydroxyapatite scaffold for engineering vascularized bone tissue using human osteoblasts: *In vitro* and *in vivo* studies. J Biomater Appl 2014; 29(5): 748-60.
[http://dx.doi.org/10.1177/0885328214544769] [PMID: 25062670]

[46] Li C, Jiang C, Deng Y, *et al.* RhBMP-2 loaded 3D-printed mesoporous silica/calcium phosphate cement porous scaffolds with enhanced vascularization and osteogenesis properties. Sci Rep 2017; 7: 41331.
[http://dx.doi.org/10.1038/srep41331] [PMID: 28128363]

[47] Bertassoni LE, Cecconi M, Manoharan V, *et al.* Hydrogel bioprinted microchannel networks for vascularization of tissue engineering constructs. Lab Chip 2014; 14(13): 2202-11.
[http://dx.doi.org/10.1039/C4LC00030G] [PMID: 24860845]

[48] Lin Y-H, Chuang T-Y, Chiang W-H, *et al.* The synergistic effects of graphene-contained 3D-printed calcium silicate/poly-ε-caprolactone scaffolds promote FGFR-induced osteogenic/angiogenic differentiation of mesenchymal stem cells. Mater Sci Eng C 2019; 104: 109887.
[http://dx.doi.org/10.1016/j.msec.2019.109887] [PMID: 31500024]

[49] Wang H-J, Huang J-C, Hou L, Miyazawa T, Wang J-Y. Prolongation of the degradation period and improvement of the angiogenesis of zein porous scaffolds *in vivo*. J Mater Sci Mater Med 2016; 27(5): 92.
[http://dx.doi.org/10.1007/s10856-016-5697-2] [PMID: 26979976]

[50] O'BRIEN FERIGAL, *et al.* Incorporation of fibrin into a collagen-glycosaminoglycan matrix results in a scaffold with improved mechanical properties and enhanced capacity to resist cell-mediated contraction [J]. Acta Biomater 2015; 8: 022.

[51] Brougham CM, Levingstone TJ, Jockenhoevel S, Flanagan TC, O'Brien FJ. Incorporation of fibrin into a collagen-glycosaminoglycan matrix results in a scaffold with improved mechanical properties and enhanced capacity to resist cell-mediated contraction. Acta Biomater 2015; 26: 205-14.
[http://dx.doi.org/10.1016/j.actbio.2015.08.022] [PMID: 26297884]

[52] Gálvez-Montón C, Fernandez-Figueras MT, Martí M, *et al.* Neoinnervation and neovascularization of acellular pericardial-derived scaffolds in myocardial infarcts. Stem Cell Res Ther 2015; 6(1): 108.
[http://dx.doi.org/10.1186/s13287-015-0101-6] [PMID: 26205795]

[53] Iyyanki TS, Dunne LW, Zhang Q, Hubenak J, Turza KC, Butler CE. Adipose-derived stem-cell-seeded non-cross-linked porcine acellular dermal matrix increases cellular infiltration, vascular infiltration, and mechanical strength of ventral hernia repairs. Tissue Eng Part A 2015; 21(3-4): 475-85.
[http://dx.doi.org/10.1089/ten.tea.2014.0235] [PMID: 25156009]

[54] Sun W, Motta A, Shi Y, *et al.* Co-culture of outgrowth endothelial cells with human mesenchymal stem cells in silk fibroin hydrogels promotes angiogenesis. Biomed Mater 2016; 11(3): 035009.
[http://dx.doi.org/10.1088/1748-6041/11/3/035009] [PMID: 27271291]

[55] Buitinga M, Janeczek Portalska K, Cornelissen D-J, *et al.* Coculturing human islets with

proangiogenic support cells to improve islet revascularization at the subcutaneous transplantation site. Tissue Eng Part A 2016; 22(3-4): 375-85.
[http://dx.doi.org/10.1089/ten.tea.2015.0317] [PMID: 26871862]

[56]   Zigdon-Giladi H, Michaeli-Geller G, Bick T, Lewinson D, Machtei EE. Human blood-derived endothelial progenitor cells augment vasculogenesis and osteogenesis. J Clin Periodontol 2015; 42(1): 89-95.
[http://dx.doi.org/10.1111/jcpe.12325] [PMID: 25361474]

[57]   Laschke MW, Grässer C, Kleer S, *et al.* Adipose tissue-derived microvascular fragments from aged donors exhibit an impaired vascularisation capacity. Eur Cell Mater 2014; 28: 287-98.
[http://dx.doi.org/10.22203/eCM.v028a20] [PMID: 25340807]

[58]   Chen W, Thein-Han W, Weir MD, Chen Q, Xu HH. Prevascularization of biofunctional calcium phosphate cement for dental and craniofacial repairs. Dent Mater 2014; 30(5): 535-44.
[http://dx.doi.org/10.1016/j.dental.2014.02.007] [PMID: 24731858]

[59]   Nör JE, Peters MC, Christensen JB, *et al.* Engineering and characterization of functional human microvessels in immunodeficient mice. Lab Invest 2001; 81(4): 453-63.
[http://dx.doi.org/10.1038/labinvest.3780253] [PMID: 11304564]

[60]   Sakaguchi K, Shimizu T, Horaguchi S, *et al. in vitro* engineering of vascularized tissue surrogates. Sci Rep 2013; 3: 1316.
[http://dx.doi.org/10.1038/srep01316] [PMID: 23419835]

[61]   Duttenhoefer F, Lara de Freitas R, Meury T, *et al.* 3D scaffolds co-seeded with human endothelial progenitor and mesenchymal stem cells: evidence of prevascularisation within 7 days. Eur Cells Mat 2013;26:64-5.

[62]   Guerrero J, Catros S, Derkaoui S-M, *et al.* Cell interactions between human progenitor-derived endothelial cells and human mesenchymal stem cells in a three-dimensional macroporous polysaccharide-based scaffold promote osteogenesis. Acta Biomater 2013; 9(9): 8200-13.
[http://dx.doi.org/10.1016/j.actbio.2013.05.025] [PMID: 23743130]

[63]   Liu J, Liu C, Sun B, *et al.* Differentiation of rabbit bone mesenchymal stem cells into endothelial cells *in vitro* and promotion of defective bone regeneration *in vivo*. Cell Biochem Biophys 2014; 68(3): 479-87.
[http://dx.doi.org/10.1007/s12013-013-9726-1] [PMID: 23943083]

[64]   Pill K, Hofmann S, Redl H, Holnthoner W. Vascularization mediated by mesenchymal stem cells from bone marrow and adipose tissue: a comparison. Cell Regen (Lond) 2015; 4(1): 8.
[http://dx.doi.org/10.1186/s13619-015-0025-8] [PMID: 26500761]

[65]   Klar AS, Güven S, Biedermann T, *et al.* Tissue-engineered dermo-epidermal skin grafts prevascularized with adipose-derived cells. Biomaterials 2014; 35(19): 5065-78.
[http://dx.doi.org/10.1016/j.biomaterials.2014.02.049] [PMID: 24680190]

[66]   Miranville A, Heeschen C, Sengenès C, Curat CA, Busse R, Bouloumié A. Improvement of postnatal neovascularization by human adipose tissue-derived stem cells. Circulation 2004; 110(3): 349-55.
[http://dx.doi.org/10.1161/01.CIR.0000135466.16823.D0] [PMID: 15238461]

[67]   Benavides OM, Quinn JP, Pok S, Petsche Connell J, Ruano R, Jacot JG. Capillary-like network formation by human amniotic fluid-derived stem cells within fibrin/poly(ethylene glycol) hydrogels. Tissue Eng Part A 2015; 21(7-8): 1185-94.
[http://dx.doi.org/10.1089/ten.tea.2014.0288] [PMID: 25517426]

[68]   Tatara AM, Kretlow JD, Spicer PP, *et al.* Autologously generated tissue-engineered bone flaps for reconstruction of large mandibular defects in an ovine model. Tissue Eng Part A 2015; 21(9-10): 1520-8.
[http://dx.doi.org/10.1089/ten.tea.2014.0426] [PMID: 25603924]

[69]   Kaempfen A, Todorov A, Güven S, *et al.* Engraftment of prevascularized, tissue engineered constructs

in a novel rabbit segmental bone defect model. Int J Mol Sci 2015; 16(6): 12616-30.
[http://dx.doi.org/10.3390/ijms160612616] [PMID: 26053395]

[70]   Ginjupalli K, Shavi GV, Averineni RK, Bhat M, Udupa N, Upadhya PN. Poly (α-hydroxy acid) based polymers: A review on material and degradation aspects. Polym Degrad Stabil 2017; 144: 520-35.
[http://dx.doi.org/10.1016/j.polymdegradstab.2017.08.024]

[71]   Chung C, Burdick JA. Engineering cartilage tissue. Adv Drug Deliv Rev 2008; 60(2): 243-62.
[http://dx.doi.org/10.1016/j.addr.2007.08.027] [PMID: 17976858]

[72]   Slaughter BV, Khurshid SS, Fisher OZ, Khademhosseini A, Peppas NA. Hydrogels in regenerative medicine. Adv Mater 2009; 21(32-33): 3307-29.
[http://dx.doi.org/10.1002/adma.200802106] [PMID: 20882499]

[73]   Mistry N, Moskow J, Shelke NB, Yadav S, Berg-Foels WSV, Kumbar SG. Chapter 5 – Bio-Instructive scaffolds for cartilage regeneration. Musculoskeletal Tissue Eng Regen Med. 2017:115-35.

[74]   Laycock B, Nikolić M, Colwell JM, *et al.* Lifetime prediction of biodegradable polymers. Prog Polym Sci 2017; 71: 144-89.
[http://dx.doi.org/10.1016/j.progpolymsci.2017.02.004]

[75]   Tesfamariam B. Bioresorbable vascular scaffolds: Biodegradation, drug delivery and vascular remodeling. Pharmacol Res 2016; 107: 163-71.
[http://dx.doi.org/10.1016/j.phrs.2016.03.020] [PMID: 27001225]

[76]   Zhu J. Bioactive modification of poly(ethylene glycol) hydrogels for tissue engineering. Biomaterials 2010; 31(17): 4639-56.
[http://dx.doi.org/10.1016/j.biomaterials.2010.02.044] [PMID: 20303169]

[77]   Shigemitsu H, Fujisaku T, Onogi S, Yoshii T, Ikeda M, Hamachi I. Preparation of supramolecular hydrogel-enzyme hybrids exhibiting biomolecule-responsive gel degradation. Nat Protoc 2016; 11(9): 1744-56.
[http://dx.doi.org/10.1038/nprot.2016.099] [PMID: 27560177]

[78]   Ifkovits JL, Burdick JA. Review: photopolymerizable and degradable biomaterials for tissue engineering applications. Tissue Eng 2007; 13(10): 2369-85.
[http://dx.doi.org/10.1089/ten.2007.0093] [PMID: 17658993]

[79]   Tajbakhsh S, Hajiali F. A comprehensive study on the fabrication and properties of biocomposites of poly(lactic acid)/ceramics for bone tissue engineering. Mater Sci Eng C 2017; 70(Pt 1): 897-912.
[http://dx.doi.org/10.1016/j.msec.2016.09.008] [PMID: 27770967]

[80]   Xinfeng C, Yong J, Rui Q, Baozhu F, Hanping L. Stimuli-Responsive Degradable Polymeric Hydrogels. Huaxue Jinzhan 2015; 27(12): 1784-98.

[81]   Gupta A, Kumar V. New emerging trends in synthetic biodegradable polymers–Polylactide: A critique. Eur Polym J 2007; 43(10): 4053-74.
[http://dx.doi.org/10.1016/j.eurpolymj.2007.06.045]

[82]   Singh A, Peppas NA. Hydrogels and scaffolds for immunomodulation. Adv Mater 2014; 26(38): 6530-41.
[http://dx.doi.org/10.1002/adma.201402105] [PMID: 25155610]

[83]   Davis KA, Anseth KS. Controlled release from crosslinked degradable networks. Crit Rev Therap Drug Carr Sys 2002; 19: 4-5.
[http://dx.doi.org/10.1615/CritRevTherDrugCarrierSyst.v19.i45.30]

[84]   Kopeček J, Yang J. Smart self-assembled hybrid hydrogel biomaterials. Angew Chem Int Ed Engl 2012; 51(30): 7396-417.
[http://dx.doi.org/10.1002/anie.201201040] [PMID: 22806947]

[85]   Yildirimer L, Seifalian AM. Three-dimensional biomaterial degradation - Material choice, design and extrinsic factor considerations. Biotechnol Adv 2014; 32(5): 984-99.

[http://dx.doi.org/10.1016/j.biotechadv.2014.04.014] [PMID: 24858478]

[86]   Shirazi RN, Ronan W, Rochev Y, McHugh P. Modelling the degradation and elastic properties of poly(lactic-co-glycolic acid) films and regular open-cell tissue engineering scaffolds. J Mech Behav Biomed Mater 2016; 54: 48-59.
[http://dx.doi.org/10.1016/j.jmbbm.2015.08.030] [PMID: 26414516]

[87]   João CF, Kullberg AT, Silva JC, Borges JP. Chitosan Inverted Colloidal Crystal scaffolds: Influence of molecular weight on structural stability. Mater Lett 2017; 193: 50-3.
[http://dx.doi.org/10.1016/j.matlet.2017.01.096]

[88]   Wan Y, Tu C, Yang J, Bei J, Wang S. Influences of ammonia plasma treatment on modifying depth and degradation of poly(L-lactide) scaffolds. Biomaterials 2006; 27(13): 2699-704.
[http://dx.doi.org/10.1016/j.biomaterials.2005.12.007] [PMID: 16412503]

[89]   Pan T, Song W, Cao X, Wang Y. 3D bioplotting of gelatin/alginate scaffolds for tissue engineering: influence of crosslinking degree and pore architecture on physicochemical properties. J Mater Sci Technol 2016; 32(9): 889-900.
[http://dx.doi.org/10.1016/j.jmst.2016.01.007]

[90]   Frueh FS, Menger MD, Lindenblatt N, Giovanoli P, Laschke MW. Current and emerging vascularization strategies in skin tissue engineering. Crit Rev Biotechnol 2017; 37(5): 613-25.
[http://dx.doi.org/10.1080/07388551.2016.1209157] [PMID: 27439727]

[91]   Phelps EA, García AJ. Engineering more than a cell: vascularization strategies in tissue engineering. Curr Opin Biotechnol 2010; 21(5): 704-9.
[http://dx.doi.org/10.1016/j.copbio.2010.06.005] [PMID: 20638268]

[92]   Zhan W, Marre D, Mitchell GM, Morrison WA, Lim SY. Tissue engineering by intrinsic vascularization in an *in vivo* tissue engineering chamber. J Visual Exp Jove. 2016;2016(111).
[http://dx.doi.org/10.3791/54099] [PMID: 27286267]

[93]   Tian T, Zhang T, Lin Y, Cai X. Vascularization in craniofacial bone tissue engineering. J Dent Res 2018; 97(9): 969-76.
[http://dx.doi.org/10.1177/0022034518767120] [PMID: 29608865]

[94]   Fu J, Wang DA. *In situ* organ-specific vascularization in tissue engineering. Trends Biotechnol 2018; 36(8): 834-49.
[http://dx.doi.org/10.1016/j.tibtech.2018.02.012] [PMID: 29555346]

[95]   Bose S, Tarafder S, Bandyopadhyay A. Effect of chemistry on osteogenesis and angiogenesis towards bone tissue engineering using 3D printed scaffolds. Ann Biomed Eng 2017; 45(1): 261-72.
[http://dx.doi.org/10.1007/s10439-016-1646-y] [PMID: 27287311]

[96]   Deng Y, Jiang C, Li C, *et al.* 3D printed scaffolds of calcium silicate-doped β-TCP synergize with co-cultured endothelial and stromal cells to promote vascularization and bone formation. Sci Rep 2017; 7(1): 5588.
[http://dx.doi.org/10.1038/s41598-017-05196-1] [PMID: 28717129]

SUBJECT INDEX

www.ingramcontent.com/pod-product-compliance
Lightning Source LLC
Chambersburg PA
CBHW041713210326
41598CB00007B/632